深度解析 Oracle
——实战与提高

◎ 付培利 主编
梁世强 徐茹 参编

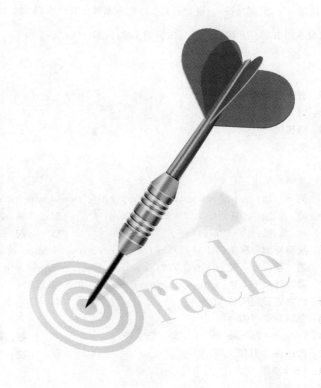

清华大学出版社
北京

内 容 简 介

本书是 Oracle 数据库大师付培利先生十几年工作经验精华的提炼。内含大量的实际工作经验,能有预见性地避免数据库问题的发生。让数据库系统一开始就完美。每个命令都可以运用到实际工作中,解决具体问题。本书所有命令,都经过作者反复测试才敢落到书上,请读者放心使用。

本书覆盖面广,从数据库基础开始,到 RAC、DataGuard、GoldenGate 的安装配置,再到较高级的性能调整、故障处理、系统容灾都有涉及,可以说适合各个学习阶段,也适合甲方在自己的数据库规划管理中,利用本书中的思想,在一些细节问题的处置上,规划调整好自己的数据库系统。数据库学习人员,对照本书中的各种实验,加强练习,注意消化和吸收,一定能有所突破,加速成长为技术专家。

本书语言独到,叙述问题一步到位,内容非常实用,可以作为普通高校计算机专业学生数据库实践教材或参考书,也适用于计算机培训班及计算机自学读者参考。

本书封面贴有清华大学出版社防伪标签,无标签者不得销售。
版权所有,侵权必究。侵权举报电话: 010-62782989　13701121933

图书在版编目(CIP)数据

深度解析 Oracle:实战与提高/付培利主编. --北京:清华大学出版社,2016
ISBN 978-7-302-44840-2

Ⅰ. ①深… Ⅱ. ①付… Ⅲ. ①关系数据库系统 Ⅳ. ①TP311.138

中国版本图书馆 CIP 数据核字(2016)第 197325 号

责任编辑:贾　斌　薛　阳
封面设计:刘　键
责任校对:胡伟民
责任印制:沈　露

出版发行:清华大学出版社
网　　址:http://www.tup.com.cn, http://www.wqbook.com
地　　址:北京清华大学学研大厦 A 座　　邮　编:100084
社 总 机:010-62770175　　邮　购:010-62786544
投稿与读者服务:010-62776969, c-service@tup.tsinghua.edu.cn
质 量 反 馈:010-62772015, zhiliang@tup.tsinghua.edu.cn
课 件 下 载:http://www.tup.com.cn, 010-62795954

印 装 者:北京密云胶印厂
经　　销:全国新华书店
开　　本:190mm×260mm　　印　张:16.25　　字　数:402 千字
版　　次:2016 年 8 月第 1 版　　印　次:2016 年 8 月第 1 次印刷
印　　数:1~3000
定　　价:45.00 元

产品编号:070502-01

北京樱溚科技有限公司
BEIJING CHERRY TIMELY RAIN TECHNOLOGY LIMITED

1. 公司简介

北京樱溚科技有限公司于 2011 年 2 月 12 日注册成立。寓意"及时雨"。樱溚的前身溯源于 1996 年 Oracle 中国的数据库调优部，2011 年正式进入 IT 系统综合服务行业。致力于为中国用户提供成熟、稳健的信息技术服务，以富有远见的科技推动工作与生活的变革。

樱溚信守以身作则、说到做到、勇于担当的准则，站在客户的立场思考问题，责任与客户保持一致。技术人员在客户现场提供贴心服务。尽心尽责的客户服务使我们拥有持续而长久的客户合作关系。

公司总经理付培利满怀信心地向您提供我们的产品和服务。我们十分乐意运用所掌握的广泛知识和技术为您服务。

2. Oracle 数据库服务事业部

（1）数据库性能优化

❶ **性能信息收集**　根据客户实际情况对客户系统各个业务高峰期的系统进行监控并收集相关数据。根据收集的系统运行数据，针对系统运行状态、运行效率、故障隐患进行评估和分析，明确定位出系统瓶颈，全面提示系统存在的各种问题、存在的风险，分析其原因。

❷ **性能评估与分析**　与客户一起收集优化前的性能指标数据，确定 BASELINE，由于优化项目的复杂性与特殊性，此阶段仅对确定的性能指标的优化效果做出简单的评估，重点对所收集的性能数据进行分析并初步确定优化重点。

❸ **制定优化方案**　优化方案的内容包含如下：对操作系统层的建议，对数据库参数的建议，平衡各模块中心数据，平衡数据库间 IO 冲突的建议，存储方面的调整方案，数据库物理层调整方案，应用层表、索引调整方案，应用层 SQL 调整方案，应用层程序调整方案。

❹ **制定应急方案**　对于可能引起系统安全问题的优化建议，在实施前，将对客户提出应急方案建议，根据不同的情况，相关厂商分别负责完成各自部分的应急方案。

应急方案包含：
- ◆ 调整期数据的备份方案；
- ◆ 调整期过程中的回退方案；
- ◆ 调整期的恢复方案。

以上方案，要根据具体的调整方案来定，如果调整方案对数据有安全影响，则制订以上方案，如果调整方案对数据没有安全影响，例如需调整程序时，则需请开发商制订应用程序备份方案。

❺ **方案测试**：根据实际情况，进行相应的优化方案测试与应急方案测试。根据方案的具体情况，如果调整属于数据库之外的其他方面，则同相应的厂商配合实施。

❻ **优化实施**：根据方案的具体情况，如调整属于数据库之外的其他方面，则同相应的厂商配合实施。

❼ **文档移交**：项目完成时将把优化服务期间形成的文档移交给客户备案。

性能优化是樱澍的竞争力所在。

（2）Trouble Shooting 紧急救援

快速提出解决方案，把问题设定为最高级别，协调公司内外资源，以解决该问题。针对数据库宕机，数据坏块，数据恢复，影响业务不能运行的产品问题进行快速解决。

（3）方案规划

系统新上线或系统升级时，樱澍对数据库系统进行规划、设计。

（4）软件产品的安装部署及测试升级

实施对于信息系统非常重要。我们以成熟的技术，把软件配置得坚如磐石。包括版本 Version 升级、更新（Release），软件修补包（Patch Fixes）及周边工作（Workarounds）。

（5）DBA 技能培训

课程包括：OCM 考题详解，常见 Case 诊断及解决，备份和恢复，数据库性能的监控和调整，运维及指标规范，DBA 基础等。

（6）热线服务

通过电话、即时通信工具和 E-mail 等形式和客户进行交流和探讨，以得到详尽的解答和技术资料等。或者通过 Internet 或拨号远程登录到机器上处理问题。

（7）健康检查服务

健康检查可以防患于未然。樱澍基于对系统的理解，全面地对主机、存储、网络、数据库、中间件和应用程序 SQL 语句等方面的运行状况进行监控检查与分析。

（8）Expert ONSITE 现场支持服务

Expert ONSITE 现场支持服务是一种灵活的现场支持服务。对于客户关键业务系统，

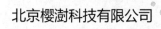

樱澍派有经验的 Oracle DBA 常住客户现场，贴心服务。

3. 代理软件硬件产品销售

北京樱澍科技有限公司充分利用自身技术团队能力和实施经验优势，为客户提供综合IT技术服务和系统集成服务。包括存储系统、系统平台集成、数据保护、灾备业务、应用与数据迁移、新机房建设与旧机房改造。

4. 联系我们

正如一个蹒跚学步的幼儿，樱澍成长的每一步都离不开您的引导和激励。请给予我们一如既往的支持。如果您有好的意见或建议；如果您对我们的工作尚有不满意之处；如果您想对樱澍有更进一步的了解……，请与我们联系，我们渴望与您交流，在交流中成长。

注册地址：北京市通州区聚富苑民族产业发展基地聚和六街二号

邮编：101105

电话：13381029910,13691321567,010-56853471

电子邮箱：fupeili@163.com 或 cherrytimerain@163.com

公司网址：http://www.cherryrain.cn/

推 荐

Oracle 数据库管理员(DBA)是 IT 从业人员中压力最大的一群人之一,他们责任重大,往往关系到数据的安全和系统的正常运行。DBA 的技能需要日常的经验积累和项目的磨炼。此书中总结了一些 DBA 在日常工作中的经验和问题的处理方法,希望对相关人员有所帮助。

刘向东(超过 15 年的 Oracle 原厂顾问
服务经验,在金融电信行业享有盛誉,代表中
国 Oracle 数据库技术最高水平)

在校大学生应该加强实践教育,走出校门后,便可以有的放矢地找工作,并很快适应工作环境。而这本书,从实践中来,计算机等专业的学生应该好好研究此书。

洪贞银教授,博士,鄂州职业大学校长

不知道"只有偏执狂才能生存"这句话是否过于夸张,我眼中的小付执着勤奋,能吃苦,有一股狠劲拼劲,又愿意以出书的方式与他人分享辛辛苦苦积累沉淀的成果,我必须给他点个赞!

神州数码技术部总经理:何晓刚

付总是实战派,十几年的 DB 售后经验,自然妙笔生花。他现在推行的是治未病技术,不让系统出问题,一开始就完美。

Oracle 公司技术顾问:于轶刚

付总与樱澍科技开创了与众不同的企业级 IT 服务模式,定能成就一帆快意江湖。

幸福人寿保险股份有限公司:赵立

数据是银行的核心元素，而数据库是数据处理最重要的部分，其运行性能的优劣，直接影响着整个银行系统的服务质量。本书从日常巡检到备份、调优，浓缩了作者十余年的 Oracle 维护技艺，从多角度分析问题，力求更好的解决途径。阅读此书，一定会令更多读者获益。

中国建设银行 DBA：胡淮杨

序

他们吓唬人

大学虽然学的是计算机,但毕业后对 Oracle 的认识几乎是零。

脱离校园的围墙和家长的屋檐开始自己谋生,总得找一个营生,还要有前景的。孙悟空学艺,被问及:"你学三十六变,还是七十二变?"答:"我学多的",还要追问:"可得长生吗?"。

对于一个北漂来说,混不好就要回老家种田,从此就没戏了。这样的例子我身边有的是,我死也不想回去。

偶然听说 Oracle DBA 工资是 IT 工程师圈里比较高的,又有声音说这个数据库技术比较难学,凭咱的智商根本学不会。就凭你能学会吗?现在看来,是根本不难,是他们吓唬人。

微观调控

中央政府实施宏观调控,而我们要做的是微观调控。把我们的数据库调整好,运行稳定流畅;把自己的家庭调理好,温馨幸福;还要把股东们的公司经营好,为股东利益负责;为客户提供最优质的产品和服务;为国家多交税。为此我感到荣幸。

善于总结

做 Oracle 数据库技术服务十几年,从一开始如履薄冰,到现在得心应手。服务过各式各样的客户,国有银行、移动联通,每秒钟上千甚至几千笔交易;也有每小时只工作半分钟的数据库。我有一个习惯,每服务一个客户,就建一个以客户单位名称为名的文件夹,记录保留相关文档,便于后续的工作;每发现并处理一个问题,遇到的新知识,记到相关的 txt 脚本里,放到名为 my script 的文件夹里,下次再遇到时方便查询。

经验是靠工作中善于总结,不断积累得来的,日积月累,自己的工作方案模板就比较齐全了。当然初学者要系统地学习基础理论知识,我初学时记的笔记就有十几本。

还是老办法

其实数据库知识是不容易过期的，Oracle 8i 的知识现在也能用得着。Oracle 9i 到 Oracle 10g，再到 Oracle 11g、Oracle 12c，新旧版本的过渡期一般为四五年，学习过的旧版本知识并不过期，只需要再学习一下新版本的新特性就行了。

去客户现场做服务发现几个问题，几经周折终于解决了，一定要记录下解决办法并留存文档。下次去客户现场，还是这些问题，还是老办法，问题以前都遇见过了。

问问题

有的学生下了课或遇见老师，总是有很多问题，有的还整理成书面的，一个一个地问，一直到搞明白。问问题的学生证明他真正地学习了，学习中发现了问题，问清楚就变成了自己的知识。有的学生却相反，从来不提问题，当你问他问题时，发现他啥也不懂。

解决问题的方法

有人说解决问题的方法是 baidu、google、metalink，就是 support.oracle.com。解决问题的方法要经过自己的测试验证和官方的支持，baidu、google 查出来的思路最好也到 metalink 上去验证一下，再一次确认可行。解决问题的方法也要靠经验的积累。

做实验

谁的生产库也不会拿来让我们当作练兵场，面临平时的学习任务，很重要的就是理论和实验相结合，做过几次实验，印象就会很深刻。现在的笔记本电脑内存又大又便宜，装几个 Linux 虚拟机做 RAC，DataGuard 都没有问题。做实验的条件应该很容易就达到了，OCM 考试就是手工操作做实验。务必多做实验。

Oracle 数据库服务工作

按工作内容分，Oracle 数据库服务工作一般包括数据库服务器系统规划、数据库安装、数据库巡检、数据迁移、生产库割接、版本升级、故障处理、性能调优、容灾、教育培训等。本书详述了所有工作内容的案例。

认证考试

考 OCA、OCP、OCM 是为了获取一份好工作，考试过程也是一个学习过程。OCA、OCP 考试准备就是背题库、背答案，技术要求不高，技术好坏和 OCP 也没什么关系。OCM 考试准备是按题库做实验，很是折磨人。水平高的人可以不用考认证。考认证可能还有其他的目的，另当别论。

欣慰

书是知识的来源之一，如果这本书中的几句话、几个命令让您印象深刻，受到启发，对您以后的工作和生活有所帮助，那我会为此感到很欣慰。

需要点拨

从农村到县城,再到省城、京城,再到国外,这个跨度还是很大的;从一个月挣几百块钱,到几千块钱,再到一个月几万块钱,这个跨度也是很大的。

在奋斗过程中,很需要高人的点拨,也就是过来人给你指指路,否则,很难突围。大学毕业刚到北京,啥也不懂。那时候只知道给领导跑前跑后,好在情商比较高,忍了不少委屈,也好在总有贵人相助,主要是对思想的影响,也可能是醍醐灌顶,明白了很多。杨总说:有些人一辈子糊涂,死也成不了明白人。

我们的任务就是帮助更多的人成为明白人。

现在社会,我们会遭受各种打击,打击只会让我们更加坚强,我们的抗打击能力也要无比地强。

苍天无眼

大学毕业,盲目闯荡一段时间之后发现自己喜欢学习技术,只有知识靠得住,看书也能看得进去。所有时间几乎除了吃饭睡觉之外就是看书,上班时间看书,下班时间也看书,坐在公交车上也看,售票员问我说公交车上还这么认真看书?我说:我要考试了。中文的看,英语的也看,看不懂就查字典。一本英文书上密密麻麻全是我写的注释。那时候认证考试得 100 分。

有一个同事说:"如果你找不到好工作,而那些整天游手好闲、无所事事、不学习的人能找到好工作,苍天无眼。"

靠山

千万不要指望这个亲戚、那个朋友。对于平民百姓来说,他们的面子大不到可以支撑你一生的幸福,可能他们也不是不想帮你。到时候别因为靠东靠西耽误了宝贵的时间,还是早早地加入到社会竞争中来才是硬道理。最后还是要靠自己赤手空拳来阅历和体验。什么东西靠得住?知识靠得住。考察好,系统地学习一门有前景的科学知识,靠得住。很多时候靠山山倒,靠水水流。

升迁

工作一段时间之后,自认有功劳,总想有一天在单位能得到升迁。可现实是:一个部门领导的位置,十几个人在排队,根本挤不动。我不得不打消在单位升迁的念头,改为自己经营公司,自己给自己升迁了。

本书特色

本书的所有命令,都经过精心的测试才落到书上,是十几年工作经验精华的提炼。本书覆盖面广,从数据库基础开始,到 RAC、DataDuard、GoldenGate 的安装配置,再到较高级的性能调整、故障处理、容灾系统都有涉及,可以说适合各个学习阶段,也适合甲方在自己的数据库规划管理中,规划调整好自己的数据库系统。

初中级学习人员,对照本书中的各种实验,加强练习,注意消化和吸收,有看不懂的地方，通过查询资料和网络搜索加深理解，经过一段时间的实践经验，一定能有所突破，加速成长为技术专家。

本书中的思路，在一些细节问题的处置上，对甲方相关工作人员很有帮助。

感　谢

感谢客户与读者选择我们的产品和服务，我们将会十分真诚地为您提供贴心服务。

感谢在我困难的时候帮助过我的朋友。

感谢恩师刘向东、于轶刚、杨磊。

感谢母校鄂州职业大学，感谢神州数码。

感谢妻子对我十几年来风风雨雨的陪伴，对家庭的辛勤付出。

感谢同仁：崔兰怀，樊志宏，何小刚，赵立，金星辰，张国刚，姚旭，桑戟楠，赵磊，林宇泽，崔鹏，蔡恒恒，张力，巩子东，卢秀红，郁超，梁世强，刘宁，董乾，王占林，殷飞，陈济民，陶鹏，王爱平，吕小京。

由于编者水平有限，书中难免存在疏漏之处，敬请各位专家和读者批评指正，作者的E-mail：fupeili@163.com，QQ：535000791。感谢您使用本教材，期待本教材能成为您的良师益友。

付培利

2016 年 6 月

目录

第1章 数据库体系结构 / 1

1.1 Oracle 历史 / 1
1.2 Oracle 体系结构 / 2
1.3 Oracle 10g/11gRAC 体系结构 / 5

第2章 数据库巡检 / 9

2.1 数据库巡检关注的内容 / 9
2.2 数据库巡检过程 / 10
2.3 快速巡检脚本/方法 / 25

第3章 数据库系统的规划 / 34

3.1 概述 / 34
3.2 规划项目时的考虑因素 / 34
3.3 项目的规划 / 37

第4章 安装 RAC 数据库 / 42

4.1 安装 RAC 前的准备工作 / 42
4.2 删除 Linux 下的 Oracle 数据库 / 43
4.3 RAC 安装过程 / 44

第5章 数据库的备份和恢复 / 77

5.1 备份和恢复概述 / 77
5.2 RMAN 备份的特点 / 78
　　5.2.1 RMAN 备份的优点 / 78
　　5.2.2 RMAN 备份的缺点 / 79
5.3 RMAN 的配置 / 79
　　5.3.1 非归档切换到归档日志模式 / 79
　　5.3.2 RMAN 的配置 / 80
5.4 数据库全备 / 83
5.5 备份部分内容 / 83

5.6 增量备份 / 84
5.7 查看备份情况 / 84
5.8 数据库恢复 / 85

第6章 数据库故障处理 / 88

6.1 故障处理概述 / 88
6.2 最常见的一些故障 / 88
6.3 案例1 存储损坏 / 89
6.4 案例2 绑定变量问题 / 90
6.5 经典案例3 数据库无法启动 / 92
6.6 案例4 OEM bug / 95
6.7 案例5 网络故障 / 95
6.8 案例6 数据库版本等问题 / 99
  6.8.1 数据库ysms的首次优化方案 / 99
  6.8.2 数据库ysms的第二次优化方案 / 101
6.9 案例7 数据库改造方案(简版) / 103
6.10 案例8 数据库参数设置问题 / 104
6.11 案例9 回闪区的限额被占满 / 109

第7章 数据库调优艺术 / 111

7.1 性能问题存在的背景 / 111
7.2 收集和了解哪些信息 / 111
7.3 调优的依据和手段 / 112
7.4 性能优化的定义和范围 / 122
7.5 性能优化的目标 / 123
7.6 深入研究数据库系统的五大资源 / 123
7.7 性能优化需要考虑的问题 / 123
7.8 风险防范措施 / 124
7.9 数据库优化结果的保持 / 124
7.10 客户收益 / 125

第8章 Oracle数据库的迁移 / 126

8.1 概述 / 126
8.2 常用的数据库迁移方法 / 126
8.3 迁移方案一 / 127
  8.3.1 概述 / 127

8.3.2 编写目的 / 127
8.3.3 迁移时间 / 127
8.3.4 数据库迁移规划方案 / 127
8.3.5 迁移前的准备 / 127
8.3.6 迁移过程 / 128

第9章 OCM考试练习实验 / 130

9.1 手工建库 / 130
9.2 数据库设置和undo管理 / 132
9.3 创建listener / 133
9.4 共享服务配置 / 134
9.5 客户端网络服务配置 / 134
9.6 表空间的创建和配置 / 136
9.7 日志文件管理 / 137
9.8 创建模式（Schema） / 138
9.9 模式的统计信息和参数文件配置 / 139
9.10 数据库的备份和高可用 / 139
9.11 创建一个数据库 / 139
9.12 安装grid control / 140
9.13 使用Grid Control / 142
9.14 实现调度器（Schedules）和定时任务（Jobs） / 143
9.15 创建一个RMAN恢复目录（Catalog） / 145
9.16 使用RMAN / 145
9.17 回闪数据库 / 146
9.18 实体化视图 / 149
9.19 手工刷新实体化视图 / 149
9.20 外部表的使用（Oracle_Loader External Tables） / 150
9.21 传输表空间（Transportable Tablespace） / 152
9.22 创建一个附加的数据缓冲区 / 154
9.23 创建大文件表空间 / 154
9.24 管理用户数据 / 154
9.25 分区表 / 156
9.26 细粒度审计 / 159
9.27 配置资源管理（使用Grid Control操作） / 161
9.28 管理实例的内存结构 / 162
9.29 管理对象的性能 / 162

9.30　statspack 报告　/ 165
9.31　安装 RAC　/ 166
9.32　配置 DataGuard　/ 167

第 10 章　数据库的升级和补丁　/ 168

10.1　版本补丁概况　/ 168
10.2　补丁的分类　/ 169
10.3　升级前的准备工作　/ 170
10.4　版本升级　/ 170
　　10.4.1　grid 打补丁 11.2.0.4.0 升级至 11.2.0.4.7　/ 170
　　10.4.2　数据库打补丁 11.2.0.4.0 升级至 11.2.0.4.7　/ 173
10.5　版本升级总结　/ 176

第 11 章　ASM　/ 177

11.1　ASM 产生的背景　/ 177
11.2　ASM 的优势和特点　/ 177
11.3　10g ASM 和 11g ASM　/ 178
11.4　ASM 双存储实验　/ 178
11.5　ASM 换存储实验（加盘减盘）　/ 180
11.6　与 ASM 相关的命令和视图　/ 182

第 12 章　DataGuard　/ 186

12.1　DataGuard 简介　/ 186
12.2　配置一个最常用的物理 DataGuard　/ 187
　　12.2.1　将主库改为归档模式　/ 187
　　12.2.2　将主库改为强制归档　/ 189
　　12.2.3　配置主库的 tnsnames.ora　/ 189
　　12.2.4　配置主库的参数　/ 190
　　12.2.5　备份主库、备份控制文件　/ 191
　　12.2.6　拷贝所需的文件　/ 192
　　12.2.7　启动备库的 listener　/ 193
　　12.2.8　修改备库的 pfile　/ 193
　　12.2.9　恢复控制文件　/ 195
　　12.2.10　恢复备库　/ 195
　　12.2.11　启动数据库　/ 196
　　12.2.12　验证 DataGuard 两边是否同步　/ 196

12.2.13 DataGuard 相关的几个重要视图 / 196
12.3 主备切换 switch over / 196
12.4 FAILEOVER 切换实验 / 198

第 13 章　Oracle GoldenGate 实施参考 / 199

13.1 概述 / 199
13.2 深入了解 GoldenGate / 199
13.3 配置一个常用的 GoldenGate / 200

第 14 章　常用 Oracle 工具在实际生产中的使用案例 / 209

14.1 10053 事件介绍及使用案例 / 209
14.2 10046 事件介绍及使用案例 / 217
14.3 SQL 优化利器之 SQL Profile 使用案例 / 223

第 15 章　Oracle 12c 介绍 / 235

15.1 Oracle Database 12c 简介 / 235
15.2 Oracle 12c 体系结构 / 235
15.3 Oracle 12c 新特性介绍 / 236

第 1 章 数据库体系结构

学习 Oracle 数据库,一开始就是学习 Oracle 历史,体系结构。我们先来看一下 Oracle 的历史,再来看它的体系结构。

1.1 Oracle 历史

Oracle 一开始是从军用开始,军方要研发一个存放数据的软件,名为 Oracle。埃里森是其中的程序员。

后来他觉得民用也会有市场,于是 1977 年埃里森和同事成立软件开发实验室,开发出了新版本的 Oracle 数据库。

1978 年公司迁往硅谷,更名为 RSI 关系软件公司,开发出了商用 Oracle 数据库。这个数据库整合了查询功能、表连接等其他一些特性。美国的中央情报局想买一套有这样功能的软件,先找 IBM 买,但 IBM 没有这种软件。于是联系了 RSI,Oracle 也就有了第一个客户。

经过几次的改名,1982 年最终把公司名改为 Oracle。

Oracle 1989 年进入中国,1991 年在北京成立甲骨文中国公司。

后来中国电信行业采用了 Oracle 数据库,一下子就火起来了。再向后,金融行业、政府机关也采用 Oracle 数据库,自此 Oracle 在中国就发展起来了。

Oracle 理念是最初版本让大家随便用,软件随便下载,很容易就能得到它的安装介质,练习实验很方便,等大家用得多了,用的顺手了,再卖 license 赚钱。

但软件补丁得要服务账号才能下载到,下载补丁把数据库升级至最稳定的版本之后,数据库产品才可以高枕无忧,否则使用最初发行的版本,bug 多得像天上的星星,系统压力一大,根本无法使用。这种方法,也可以控制一部分盗版。

让 20% 的人付 80% 的钱,二八定律。Oracle 软件授权卖得很贵,按 CPU 或按用户数卖,让大用户、大公司付高昂的 license 费用。同时也有部门面向小用户,如个人学习培训等。

1.2　Oracle 体系结构

Oracle 体系结构如图 1-1 所示。RAC 和 12c 的体系结构就是在这一张图上有所扩展，其根本还是这一张图。只要我们把这一张图搞明白了，事半功倍。

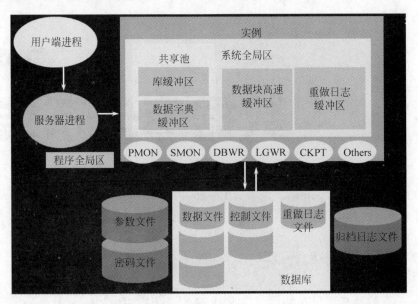

图　1-1

我们按照从左到右、从上到下的顺序，详细介绍这一张图。

用户端进程（User Process）：相当于用户台式机或笔记本电脑或中间件服务器产生的一个进程，要和数据库服务器产生连接，形成一个通道。这个在客户端的进程叫用户进程。用户进程不直接和数据库交互，它通过和服务器进程建立的 session 与数据库交互。

服务器进程（Server Process）：在服务器端，和用户进程进行对话的进程叫服务器进程。在 Linux，UNIX 平台上，服务器进程有很多，共享服务模式下，和用户端进程是一一对应的关系。一个服务员专门为一个顾客服务，所以服务质量比较好，速度比较快。在 Windows 环境下，是单进程分出多个线程，线程没有进程速度快，内存利用不如进程使用得充分，因此核心库在 Windows 下较少。

会话（Session）：用户端进程和服务器进程建立的这个通路，就叫一个会话，像水管，负责来回运送用户端和服务器端的数据。不管是进程还是会话，都是肉眼看不到的，我们可以用命令查看它们，它们的状态、参数、说明都可以用命令查询到。

程序全局区（Process Global Area，PGA）：进程专用的内存区域，不共享。用来排序的，如果数据库的排序工作很多在磁盘上完成，那么证明 PGA 设小了。还有权限控制的功能。一般设置几 GB 十几 GB，具体看数据库排序工作的多少。如果没有磁盘排序，就证明够用。在实例启动时分配，但不是按参数设置的大小分配的，是按实际需求的大小分配的。

实例（Instance）：实例不叫数据库，实例是由一大块内存区域和后台进程组成的。也可以说实例包含 SGA 和后台进程。实例的特点是运算速度快，数据交换速度快，减少 I/O 的访问,相当于货运集散地,物流中心。关闭数据库之后,实例就什么都没了。

系统全局区(System Global Area,SGA):是实例的基本组成部分,是一块内存区域,是被进程共享的,不会独占,由数据高速缓冲区、共享池、重做日志缓冲区、大池、Java 池、流池(Streams Pool)组成。最好手工管理其大小,压力较小的情况下,自动管理也可以。

共享池(Shared Pool):共享池存放最近最常使用的 SQL 语句和最近最常使用的数据字典信息。当然常用的程序包,存储过程也可以在里面,还可以手工 keep 到里面,不出来。

库缓冲区(Library Cache):存放最近最常使用的 SQL 语句,被解析的结果也存放在里面,如果使用了绑定变量,下一次再执行同样的 SQL 时,就不需要硬解析了,直接使用解析好的执行计划,省去了硬解析过程,节约 CPU 等资源。

数据字典缓冲区(Data Dictionary Cache):存放最近最常使用的数据库定义,数据库自身的一些表,也包括对应用用户表、列、权限、索引及其他对象的定义。

数据块高速缓冲区(Database Buffer Cache):是 instance 中最大的一块内存区域,内存手工管理的情况下,由参数 db_cache_size 来设置。它的大小直接影响 I/O 的多少,所以我们应该认真研究 db_cache_size 的设置。

重做日志缓冲区(Redo Log Buffer):相当于一个漏斗,大小很小,由 LGWR 进程写入 redo log。每次写的数据量比较小,但写得很频繁,所以就要求 redo log 必须放在速度快的磁盘上。大小由参数 log_buffer 来设置。

后台进程:实例的组成部分,相当于一个个的水管,说它是搬运工也很贴切,保持物理数据库和内存结构的关系,分为必需的后台进程和可选的后台进程。必需的后台进程是 DBWR、PMON、CKPT、LGWR、SMON、RECO 六个。可选的后台进程有 ARCn、LMON、Snnn、QMNn、LMDn、CJQ0、Pnnn、LCKn、Dnnn。

数据库写进程(Database Writer Process,DBWR):将 buffer cache 中的脏块写入数据文件,脏块指的是被修改的数据块。遵循 LRU 算法,写最近最少使用的块。在以下情况下开始写:❶执行检查点;❷脏块数达到上限;❸缓存没有自由空间;❹超时;❺连接 RAC 要求;❻表空间 offline;❼表空间 read only;❽清空和删除表;❾表空间开始备份。

进程监控进程(Process Monitor,PMON):有一种鱼叫清道夫,清除水中的垃圾,PMON 就是这种进程。清除失效的用户进程、释放用户进程所用的资源、将未完成的事务回滚、释放锁、释放其他资源、重启死的 dispathchs。

检查点进程(Checkpoint Process,CKPT):更新数据文件头,同步数据文件、日志文件和控制文件。给 DBWR 发信号,修改控制文件。

日志写进程(Log Writer,LGWR):将 log buffer 中的数据写入 redo logfile。在以下情况下开始写:❶提交时;❷每隔三秒钟;❸在 DBWR 写之前;❹达到阈值。日志写进程的触发条件根据版本的不同会有所变化,所以 log buffer 的设置也有一些变数。

系统监控进程(System Monitor,SMON):负责清理临时空间,实例恢复,收缩 undo 段。

分布式恢复进程(Distributed Recovery,RECO):自动地解决在分布式事务中悬而未决的事务。

以上这些是单机版的后台进程,RAC 的后台进程包含单机版的后台进程、RAC 的后台进程,我们在之后的章节中详细叙述。

参数文件 spfile 或 pfile:两种参数文件可以相互转换,系统默认使用 spfile,如果没有 spfile,再去找 pfile。参数文件的位置 UNIX、Linux 系统在 ORACLE_HOME/dbs 下,

Windows 操作系统在 ORACLE_HOME/database 下,采用 ASM 管理文件时,spfile 通过 initsid.ora 指向 ASM 里面。参数文件是记录 instance 参数的文件。spfile 是二进制格式,pfile 是文本文件格式。spfile 必须使用 alter system 命令修改,有些参数可以在线修改即生效,不需要重启数据库。pfile 只能使用编辑器修改参数,必须重启数据库生效。spfile 是 pfile 的升级版,Oracle 8i 和 Oracle 8i 之前只有 pfile,Oracle 9i 才引入 spfile。

密码文件 orapwsid:位置在 ORACLE_HOME/dbs 下,Windows 操作系统在 ORACLE_HOME/database 下,和 spfile 在一个文件夹下。主要作用是认证通过或拒绝 DBA 身份的人连接到数据库里去。使用或不使用密码文件通过数据库参数 remote_login_passwordfile 来设置。以 DBA 身份登录数据库有两种方法,一是操作系统认证登录,不使用密码文件;二是密码文件认证登录数据库。启用或关闭操作系统认证使用 $ORACLE_HOME/network/admin/sqlnet.ora 文件里的参数 sqlnet.authentication_services。当此参数设置为 none 时,将关闭操作系统认证。此文件和以上设置是提高 Oracle 数据库安全的一种方法。

数据库 database:从专业的角度讲,也就是业内人士的称谓,数据库只包含三种文件即数据文件,控制文件,日志文件,其他所有文件都不叫数据库。这三种数据文件缺一不可,缺少任何一种文件数据库都无法正常运行。组成一个数据库至少需要四个文件,即一个数据文件,一个控制文件,两个日志文件。

数据文件 datafile:存放数据对象的操作系统物理文件,是数据库中最大的一种文件。我们的表、索引、数据字典、程序等对象都在数据文件中存放。相当于一个仓库。一个数据文件只能属于一个表空间,一个表空间可以包含一个或多个数据文件。数据文件包含段、区、块。段、区、块的设置也影响数据库的性能。如我们 OLAP 的数据库,可以使用 32KB 的块,提高读取的效率。

控制文件 controlfile:控制文件是一个比较小的二进制文件,大小一般为 10MB 左右,大小基本不变。存放数据库的物理结构,如数据库名、数据文件在什么地方、日志文件的名称和位置、备份信息、检查点信息等。控制文件的位置在参数文件中记录。我们一般要复用两份或三份来保护控制文件。控制文件类似于书的目录或家里的管家。

重做日志文件 redolog:相当于录影带,记录数据文件所有的变化。如果数据文件丢失,我们只需要有一份全备,再加上归档日志和在线日志,就可以做数据库的全恢复,做到一条数据也不丢。日志文件大小不变,改变它的大小只能添加新的删除旧的。它的写速度对性能的影响非常大。因为在线日志文件的大小不变,在归档模式下,写满之后,就自动 cp 一份生成归档日志文件。

归档日志文件 archivelog:归档日志是重做日志的 cope。大小等于或小于重做日志文件。它的主要作用是用于恢复数据库。数据库恢复的时候必须有备份,归档日志才有意义,没有备份,光有归档日志没用,由此看出数据库全备的重要性。归档日志文件必须定期删除,否则早晚会把空间撑满,把数据库 hang 住,一般在做完备份后删除,具体的删除策略要结合备份策略来制订。

SQL 执行流程:终端用户单击一下鼠标或执行一个动作,可能产生一条 SQL 语句或一段 PLSQL 程序,从此这条 SQL 语句就开始了一个"漫长的"执行过程,我们假设用户要执行一个查询语句。用户进程 user process 首先要通过 listener 在验证用户名和密码无误之后和 server process 建立一个 session,通过这个 session 将 SQL 语句放入共享池中的

library cache，先从中找有没有这条语句的执行计划，如果没有则要硬解析生成执行计划，有了执行计划就知道了怎么执行这条 SQL 语句，将在 buffer cache 中查找有没有想要的数据，如果有直接返回给用户，如果没有则要通过物理 I/O 在数据文件中读取，读出来之后返回给用户，这条查询语句就执行完了。

1.3　Oracle 10g/11gRAC 体系结构

　　在 Oracle 12c 之前，一个实例只能打开一个数据库，一个数据库可以被多个实例打开。一个数据库一个实例就是单机版数据库，一个数据库多个实例就是 RAC。RAC 一般有两个节点，也可以做成 4 节点、6 节点、8 节点，单数节点也可以，但不建议使用太多的节点，节点多出故障的几率会增加，管理维护成本也会增加。在本书中所涉及的 RAC 都是两个节点。和单机版功能相同的组件我们就不叙述了，我们只讲 RAC 独有的组件。

　　RAC（Real Application Clusters）是真正的应用集群。其目的是为了防止主机损坏和 CPU 不够用，对于存储损坏，RAC 无能为力。RAC 一般由多台主机一台存储组成，每台主机上都有自己的监听器，监听自己的网络端口，每台主机都有自己的集群就绪服务，用于集群管理。所有的主机通过自己的操作系统访问同一个存储，共享的存储 Oracle 10g 设备可以是集群文件系统（OCFS）、ASM（自动存储管理）、裸设备（RAW）或网络区域存储（NAS）。到了 Oracle 11g 第二版，淘汰了裸设备，增强了 ASM 的功能，共享存储只能使用 ASM 和共享文件系统了。

　　在图 1-2 中没有体现 voting disk 和 OCR 盘，这两种也是非常重要的共享文件。在 Oracle 10g 时，这两种文件必须放在裸设备上或 OCFS 上，不能放在 ASM 上，到了 Oracle 11g 第二版，voting disk 和 OCR 盘可以放在 ASM 上了，大大方便了存储的划分，所以注定 Oracle 10g 成为过渡版本，Oracle 11g 第二版才更加成熟。

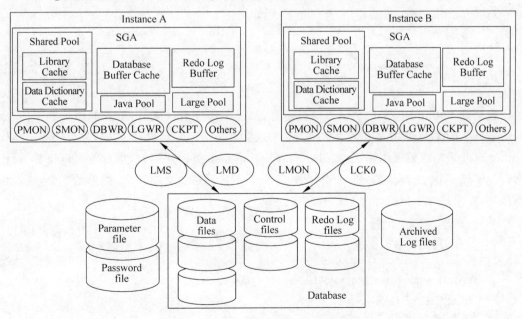

图　1-2

Oracle 集群注册表 OCR（Oracle Cluster Register）：OCR 盘记录节点配置信息，CRS 包含的资源数据库、实例、监听器、ASM、VIP 等配置信息都在 OCR 中记录。CRS 进程读取 OCR 盘来获取信息，所以 CRS 进程不启动很多命令都执行不了，srvctl、dbca、dbua、netca、crs_stat 都无法执行 RAC 操作，因为这些程序都是 CRS 的客户端。OCR 盘的大小约为 100MB。Oracle 每隔 3 个小时自动备份一次 OCR 盘，保留最后的 3 份。

```
[root@yingshu1 bin]# ./ocrconfig -showbackup
yingshu1     2015/09/17 11:31:09   /u01/11.2.0/grid/cdata/yingshu-cluster/backup00.ocr
yingshu1     2015/09/16 18:31:27   /u01/11.2.0/grid/cdata/yingshu-cluster/backup01.ocr
yingshu1     2015/09/16 14:31:26   /u01/11.2.0/grid/cdata/yingshu-cluster/backup02.ocr
yingshu1     2015/09/16 14:31:26   /u01/11.2.0/grid/cdata/yingshu-cluster/day.ocr
yingshu1     2015/09/04 13:53:34   /u01/11.2.0/grid/cdata/yingshu-cluster/week.ocr
```

可以使用以下命令对 OCR 进行恢复，恢复时 CRS 必须停止：

```
[root@yingshu1 bin]# pwd
/u01/11.2.0/grid/bin
[root@yingshu1 bin]# ./ocrconfig -restore /u01/11.2.0/grid/cdata/yingshu-cluster/backup00.ocr
```

检查当前 OCR 的状态：

```
[root@yingshu1 bin]# ./ocrcheck
Status of Oracle Cluster Registry is as follows :
         Version                  :          3
         Total space (kbytes)     :          262120
         Used space (kbytes)      :          3056
         Available space (kbytes) :          259064
         ID                       :          261064960
         Device/File Name         :          +OGOCR
                                             Device/File integrity check succeeded
                                             Device/File not configured
         Cluster registry integrity check succeeded
         Logical corruption check succeeded
```

投票盘 Voting Disk：记录节点成员信息，如包含哪些节点成员，节点的添加删除信息。voting disk 的主要作用是在发生脑裂时，决定由哪个节点获取控制权，剔除其他节点，voting disk 采用多数可用算法，超过一半的 voting disk 可以使用时，集群才可以正常运行。所以 voting disk 的个数建议是 1 个，3 个，5 个。不建议是 2 个或 4 个，如果是 2 个，一个也不能坏，如果是 4 个只能坏一个。

```
[root@yingshu1 bin]# ./crsctl query css votedisk
##  STATE    File Universal Id                File Name Disk group
--  -----    -----------------                --------- ---------
 1. ONLINE 9be94b2e72ab4f30bfff452502b37981 (/dev/raw/raw1) [OGOCR]
Located 1 voting disk(s).
[root@yingshu1 bin]#
```

RAC 的组成:

(1) cluster nodes 需要 2 到 n 个节点运行 CRS 软件及数据库软件。集群和数据库软件一般都安装在主机本地。Oracle 10g 集群和数据库软件都由 Oracle 用户来安装。Oracle 11g 引入 grid 用户,grid 只负责集群软件,Oracle 用户只负责数据库软件。

(2) 私有网络,Oracle 10g 每个主机至少需要两张网卡,每个主机至少需要设置 3 个 IP 地址。Oracle 11g 每台主机至少需要两张网卡,每台主机至少需要设置 4 个 IP 地址,比 Oracle 10g 多一个 SCAN IP 地址。两个节点的 SCAN IP 地址是一样的,为的是单一节点访问,提供服务的只需要一个 IP 就可以了。

(3) 共享存储,让所有节点都可以读写的存储设备,多路径需要做好。

(4) 对外服务的网络,我们可以把 VIP 作为对外服务的 IP 地址,也可以用 SCAN IP 作为对外服务的 IP 地址,具体看规划,甚至 public IP 也可以提供服务。

RAC 的主要组件:

(1) 服务器大于或等于两台。

(2) 操作系统推荐使用 Oracle 认证的操作系统,aix,hp-ux,Linux。不建议用 Windows。

(3) CPU/内存。建议 CPU 8 核以上,内存 32GB 以上。

(4) 本地磁盘大于 60GB。

(5) 网卡至少 2 张,建议 3 张或 4 张,两张绑成 1 张作为私有网卡内部通信使用。如果只有 3 张,那么绑内部通信的私有网络,不绑公用网络。如果 4 张网卡,可以都绑。

(6) 私有以太网,建议千 MB 以上,网线不建议两台主机直连,建议都接在交换机上以使速度更快更稳定。

(7) 共享存储,推荐使用 SAN 设备。

(8) 如果使用 SAN 设备,推荐两个 HBA 卡。

(9) 存储管理,ASM,共享文件系统,裸设备。Oracle 10g 可以使用裸设备,Oracle 11g 一般用 ASM。

(10) 集群管理软件,Oracle 10g 是 OCR 软件,Oracle 11g 是 grid 软件。

RAC 的主要后台进程如下:

```
[root@yingshu1 grid]# ps -ef|grepd.bin
root      4537     1  0 Sep16 ?        00:01:14 /u01/11.2.0/grid/bin/ohasd.bin reboot
grid      5200     1  0 Sep16 ?        00:00:18 /u01/11.2.0/grid/bin/gpnpd.bin
grid      5216     1  0 Sep16 ?        00:01:17 /u01/11.2.0/grid/bin/gipcd.bin
root      5230     1  0 Sep16 ?        00:02:19 /u01/11.2.0/grid/bin/osysmond.bin
root      5251     1  0 Sep16 ?        00:00:10 /u01/11.2.0/grid/bin/cssdmonitor
root      5293     1  0 Sep16 ?        00:00:09 /u01/11.2.0/grid/bin/cssdagent
grid      5318     1  0 Sep16 ?        00:01:28 /u01/11.2.0/grid/bin/ocssd.bin
root      5744     1  0 Sep16 ?        00:00:17 /u01/11.2.0/grid/bin/octssd.bin reboot
grid      5805     1  0 Sep16 ?        00:00:17 /u01/11.2.0/grid/bin/evmd.bin
root      7001     1  0 Sep16 ?        00:00:59 /u01/11.2.0/grid/bin/crsd.bin reboot
```

高可用服务守护进程（Oracle High Availability Services Daemon，OHASD）：Oracle 集群资源底层主持进程，是 Oracle 11g/R2 RAC 的关键进程，包含多个帮助操作集群的进程。

集群就绪服务守护进程（Cluster Ready Services Daemon，CRSD）：主要的集群进程，执行高可用恢复和管理操作。如维护 OCR，管理应用资源，在失败的时候重启应用资源。是 Oracle 10g 的核心进程，包含数据库、实例、监听、VIP，ons，gds 等资源。出故障的时候，操作系统会重启这个进程。此进程由 root 用户负责启动和关闭。这个进程不启动，srvctl 等命令就执行不了。

集群同步服务进程（Cluster Synchronization Services Daemon，CSSD）：一个 Linux，UNIX 系统中的进程，管理集群同步服务，负责各节点间的通信。节点在加入或离开集群时通知集群，发生故障时 CSSD 也会重启系统。

事件监测进程（Event Manager Daemon，EVMD）：事件管理进程，负责发布 CRS 的各种事件。

RAC 的主要警告日志：Oracle RAC 的警告日志在 grid 用户的 CRS_HOME/log/sid 下。如果发生进程无法启动，RAC 出现故障等情况，可以查看对应的某个进程的日志。本书中安装的 RAC，日志在 yingshu1 下。文件夹的名称就是进程名，进入文件夹后，即可查看这个进程的日志，由此对故障做出详细的判断。

```
[grid@yingshu1 yingshu1]$ pwd
/u01/11.2.0/grid/log/yingshu1
[grid@yingshu1 yingshu1]$ ls
acfs         acfsrep1root   agent           crflogd    cssd     diskmon   gnsd    ohasd
acfslog      acfssec        alertyingshu1.log  crfmond  ctssd    evmd      gpnpd   racg
acfsrepl     admin          client          crsd       cvu      gipcd     mdnsd   srvm
```

第 2 章 数据库巡检

2.1 数据库巡检关注的内容

数据库巡检是一个相对比较轻松的工作,是大家比较喜欢的工作,不像故障处理和割接,压力大还要加班。如果时间短,问问用户有什么性能或其他问题,我们重点先看看表空间利用率和文件系统利用率等几个要点。然后再研究一些深入的问题,给出建议,写进巡检报告里。写巡检报告有的比较长,几十页,客户又有十几套库,写文档的量比较大。我们有最简化版的巡检报告,可以省去写巡检报告长篇大论的问题,轻轻松松两三页,只写要紧的,让巡检工作成为一种享受。

下面就所用的脚本对要巡检的内容做一个介绍。以下都以 Linux 作为样例。

巡检中我们需要关注的问题主要如下:

(1) 主机配置,有多少内存;多少个 CPU,每个 CPU 多少核;存储多大,分别给谁用,RAID 情况,存储的速度。

(2) 操作系统的核心参数和信号量等设置。

(3) 网络配置,网络速度,listener 是否正常。

(4) Instance 参数配置,SGA 大小,PGA 大小,是否手工管理 SGA。

(5) 数据库配置;块大小;日志文件大小,组数;ASM、裸设备、文件系统。

(6) 数据库安全配置;用户的密码;是否授予 DBA 角色;应用用户的权限。

(7) 数据库性能。SQL 语句的响应时间,数据库整体响应时间,问题 SQL 的执行计划。

(8) 数据库的版本;小补丁情况,警告日志有无报错;报错的处理方案。

(9) 数据库的容灾和备份情况,是否使用了 DataGuard 或其他容灾方式;RMAN 备份的可用性;是否做恢复演练。

2.2 数据库巡检过程

（1）文件系统利用率 df -h。aix 命令用 df -g。hp-ux 命令用 bdf。

```
[root@yingshu ~]# df -h
Filesystem                    Size Used Avail Use% Mounted on
/dev/mapper/VolGroup00-LogVol00
                               53G  50G   3G  94% /
/dev/sda1                      99M  22M  73M  23% /boot
tmpfs                         1.8G 236M 1.6G  14% /dev/shm
none                          1.8G 104K 1.8G   1% /var/lib/xenstored
```

 文件系统/挂载点下利用率过高，建议查看被哪些文件所占用，及时清理或扩展文件系统空间。

（2）操作系统的性能 top。CPU、内存使用率，I/O 等待情况，I/O 速度。Linux 命令 sar 2 10。vmstat 2 10。aix 命令 topas 也要掌握：

```
[root@yingshu ~]# top
top - 11:41:41 up 5:45, 3 users, load average: 0.05, 0.04, 0.00
Tasks: 193 total,   1 running, 191 sleeping,   0 stopped,   1 zombie
Cpu(s): 0.0%us,  1.0%sy,  0.0%ni, 94.7%id,  3.2%wa,  0.0%hi,  0.3%si,  0.8%st
Mem:   3685376k total,  3178520k used,   506856k free,    43112k buffers
Swap:  6094840k total,      232k used,  6094608k free,  2516776k cached

  PID USER      PR  NI  VIRT  RES  SHR S %CPU %MEM    TIME+  COMMAND
 7804 root      15   0     0    0    0 S  1.7  0.0   0:04.65 pdflush
    1 root      15   0 10368  636  544 S  0.0  0.0   0:01.53 init
```

（3）虚拟内存的使用情况：

```
[root@yingshu ~]# vmstat 2 10
procs -------memory------- ---swap-- ----io---- --system-- ----cpu----
 r  b   swpd   free   buff  cache   si   so    bi    bo   in   cs us sy id wa st
 2  4 126104  56952   6528 2906768    0    1   215   705  173  318  2  4 86  8  0
 0  8 126096  55904   6532 2906980   16    0   102 44820  157  272  2 31  4 61  1
 3  7 126096  57648   6536 2907028    0    0     4 24594  183  266  0 30  0 68  1
 1  3 126096  58964   6536 2907052    0    0     0 32006  158  250  4 48  5 42  1
 1  4 126096  61568   6536 2907068    0    0     0 23028  166  251  2 43  8 45  1
 1  0 126092  63444   6552 2907056    0    0     0   506  168  298  0 49 16 35  1
```

 b 表示被阻塞的进程，如果大于 CPU 核数的两倍，则说明系统的性能较差。

(4) 系统资源的使用情况,也可以配合-d查看硬盘使用情况。

```
[root@yingshu ~]# sar 2 10
Linux 2.6.18-308.el5xen (yingshu)    09/24/2014

12:09:14 PM     CPU     %user    %nice    %system    %iowait    %steal    %idle
12:09:17 PM     all     1.56     0.00     72.57      21.60      1.17      3.11
12:09:19 PM     all     5.97     0.00     45.02      38.81      1.24      8.96
12:09:21 PM     all     1.24     0.00     60.40      34.90      0.99      2.48
12:09:23 PM     all     0.24     0.00     31.54      58.92      1.96      7.33
12:09:25 PM     all     6.63     0.00     43.98      41.03      1.47      6.88
12:09:27 PM     all     1.01     0.00     31.39      57.47      1.52      8.61
12:09:29 PM     all     3.47     0.00     41.19      49.63      1.49      4.22
12:09:31 PM     all     2.21     0.00     36.03      50.00      2.21      9.56
```

建议：本系统 I/O 等待较高,建议关注消耗 I/O 的进程和磁盘的读写速度。

(5) 数据库的基本信息

可以使用以下脚本巡检:

```
select name as "db name", instance_name, open_mode, log_mode, status, host_name from v$database, v$instance;

set head off
select 'the count of cpu: ' || value from v$parameter where name = 'CPU_COUNT';

select name from v$datafile where rownum < 2;

select 'db component version: ' || product || ' ' || version from product_component_version where upper(substr(product,1,3)) in ('ORA','TNS');

select 'sum of datafile size(g): ' || trunc(sum(bytes)/1024/1024/1024) || 'g' as sum_g from dba_data_files;
```

第一句查询：数据库名称、实例名称、开启状态、是否归档、启动状态、主机名称。

第二句查询：CPU 核数。

第三句查询：数据文件在什么地方,是裸设备、文件系统或 ASM。

第四句查询：数据库版本。

第五句查询：数据文件一共多大。一般可以认为是数据库的总大小,当然还要加上 redo 和控制文件所占的空间。

(6) 数据库设置

```
show sga;

select 'db_cache_size(m):' || value/1024/1024 from v$parameter where name = 'db_cache_size';

select 'log_buffer(k):' || value/1024 from v$parameter where name = 'log_buffer';
```

```sql
select 'shared_pool(m):' || value/1024/1024 from v$parameter where name = 'shared_pool_size';

select 'java_pool_size(m):' || value/1024/1024 from v$parameter where name = 'java_pool_size';

select 'large_pool_size(m):' || value/1024/1024 from v$parameter where name = 'large_pool_size';

select 'db_block_size(k):' || value/1024 from v$parameter where name = 'db_block_size';

select 'pga(m):' || value/1024/1024 from v$parameter where name = 'pga_aggregate_target';

select 'the count of tablespaces:    ' || count(*) from v$tablespace;

select 'the count of datafiles:    ' || count(*) from v$datafile;

select 'the count of controlfiles:    ' || count(*) from v$controlfile;
```

以上语句分别查询的是：❶sga 大小；❷数据高速缓冲区大小；❸日志缓冲区大小；❹共享池大小；❺Java 池大小；❻大池大小；❼数据块大小；❽程序全局区大小；❾表空间总个数；❿数据文件总个数；⓫控制文件总个数。

对于 Oracle 10g/11g，Oracle 可以自动管理内存大小，Oracle 10g 只需要设置一个 sga_target 和 pga_target Oracle 便可以自己管理内存的大小。Oracle 11g 新增了两个内存参数，即 memory_max_target 和 memory_target。sga 和 pga 在一起可以动态管理了。

（7）日志文件详情

```sql
select 'the size of redolog (m):    ' || bytes/1024/1024 from v$log;

select 'redo group: ',vl.* from v$log vl;

select 'redo member: ' from dual;
column member format a50
select * from v$logfile;

select 'yesterday redo average switch frequency(min) : ' || round(period/y.cnt,2) from
(select sum((a.first_time - b.first_time) * 24 * 60) as period
from v$log_history a, v$log_history b
where a.recid = b.recid + 1
and a.first_time > trunc(sysdate - 1) + 10/24
and a.first_time < trunc(sysdate - 1) + 16/24) x,
(select count(*) as cnt from v$log_history
where first_time > trunc(sysdate - 1) + 10/24
and first_time < trunc(sysdate - 1) + 16/24) y;

select rpad('arch info',100,'-') from dual;
select 'log_mode: ' || log_mode from v$database;
show parameter log_archive_dest
```

❶查询日志大小与组数；❷日志组的详情；❸日志 member 详情；❹昨天日志平均切换时间；❺归档信息；❻归档日志的路径。

(8) 许可限制

```
select sessions_max,sessions_warning,sessions_current,sessions_highwater,users_max from v
$license;
SESSIONS_MAX SESSIONS_WARNING SESSIONS_CURRENT SESSIONS_HIGHWATER USERS_MAX
------------ ---------------- ---------------- ------------------ ---------
           0                0                2                  9         0
```

这个视图包括许可限制的信息。

(9) 安装配置选件

```
set pagesize 500
column parameter format a40
column value format a30
set head on
select * from v$option;
PARAMETER                                VALUE
---------------------------------------- ------------------------------
Partitioning                             TRUE
Real Application Clusters                FALSE
Parallel backup and recovery             TRUE
Online Index Build                       TRUE
Automatic Storage Management             FALSE
Oracle Data Guard                        TRUE
Oracle Label Security                    FALSE
Advanced Compression                     TRUE
```

(10) 默认的参数和非默认的参数

```
column name format a50
column value format a50
select rpad('default parameters',50,'-') from dual;
set head on
select num,name,value from v$parameter where isdefault = 'TRUE';
set head off

select rpad('no default parameters',50,'-') from dual;
set head on
select num,name,value from v$parameter where isdefault = 'FALSE';
     NUM NAME                                          VALUE
-------- --------------------------------------------- -----------------
      39 processes                                     1500
     781 memory_target                                 1073741824
     830 db_block_size                                 8192
     982 compatible                                    11.2.0.4.0
    1501 undo_tablespace                               UNDOTBS1
    1791 remote_login_passwordfile                     EXCLUSIVE
    1804 db_domain
    1997 audit_trail                                   DB
```

```
2020 db_name                           yingshud
2021 db_unique_name                    yingshudb
2022 open_cursors                      3000
2874 diagnostic_dest                   /u01/app/oracle
```

以上是非默认参数。建议查看设置的大小是否满足生产需求。

（11）内存参数设置

```
show sga;
select 'db_cache_size(m):'   || value/1024/1024 from v$parameter where name = 'db_cache_size';
select 'log_buffer(k):'      || value/1024 from v$parameter where name = 'log_buffer';
select 'shared_pool(m):'     || value/1024/1024 from v$parameter where name = 'shared_pool_size';
select 'java_pool_size(m):'  || value/1024/1024 from v$parameter where name = 'java_pool_size';
select 'large_pool_size(m):' || value/1024/1024 from v$parameter where name = 'large_pool_size';
```

❶查询 sga 的大小；❷db_cache_size 的大小；❸log_buffer 的大小；❹共享池的大小；❺Java 池的大小；❻大池的大小。如果设置了 sga_target 或 memory_target，则❷～❻的设置不起作用。

（12）查询 sga 内部详细分配信息

```
select pool,name,bytes/1024/1024 as "size(m)" from v$sgastat where name like '%free%';
```

包括共享池、大池、Java 池、流池。如果空闲空间较多，则不需要增大 sga。

（13）数据库失效对象的查询

```
select rpad('invalid objects',100,'-') from dual;
column owner format a15
column object_name format a25
column object_type format a15
column status format a10
select owner,object_name,object_type,status from dba_objects where status = 'INVALID';
INVALID OBJECTS----------------------------------------------------------------

SYSTEM        ORA$_SYS_REP_AUTH              PROCEDURE          INVALID
SYSTEM        DBMS_REPCAT_AUTH               PACKAGE            INVALID
SYSTEM        DEF$_PROPAGATOR_TRIG           TRIGGER            INVALID
SYS           DBMS_CUBE_EXP                  PACKAGE BODY       INVALID
SYS           AWM_CREATEXDSFOLDER            FUNCTION           INVALID
YINGS         PRO_TJ_KPD_XXFPDY              PROCEDURE          INVALID
YINGS         PRO_TJ_KPFWQ_XXFPDY            PROCEDURE          INVALID
YINGS         PRO_TJ_NSR_XXFPDY_2013111      PROCEDURE          INVALID
```

 关注失效对象是否被使用，如果不被使用，则建议删除。对于应用使用的失效对象，建议重新编译。在数据库升级或迁移前后，对比失效对象的个数，判断带来的影响。

(14) 是否有 offline 的数据文件

```
select rpad('datafile is offline ? ',100,'-') from dual;
set head on
select file_name from dba_data_files where status = 'OFFLINE';
```

如果存在 offline 的数据文件,归档模式下,归档日志还在,直接做介质恢复处理,然后将数据文件 online 就可以了。如果归档日志已删除,则只能恢复整个数据库才能将此数据文件 online。

(15) 是否存在失效的约束和触发器

```
select rpad('invalid constraints ',100,'-') from dual;
select owner, constraint_name, table_name, constraint_type, status from dba_constraints where status = 'DISABLED';
select rpad('invalid triggers ',100,'-') from dual;
set head on
select owner, trigger_name, table_name, status from dba_triggers where status = 'DISABLED';
```

如果想让无效的触发器生效,则使用语句:

```
alter tigger trigger_name enable;
```

如果想让约束失效或生效,则使用:

```
alter table table_name disable constraint constraint_name;
alter table table_name enable constraint constraint_name;
```

(16) 数据文件的状态

```
select rpad('datafile info',100,'-') from dual;
column file_name format a50
column tablespace_name format a30
set head on
select file_id, file_name, tablespace_name, bytes/1024/1024 as "size(m)", status, autoextensible from dba_data_files;
```

文件 id、数据文件名称和路径、表空间名、数据文件大小、状态、是否自动扩展。

> **建议**:数据文件最好不要自动扩展,自动扩展会导致文件大小不一,某个文件异常过大,显得杂乱没有规划,自动扩展减少了维护量,但没人维护的数据库比较危险。

(17) 临时文件的状态

```
select rpad('tmpefile info',100,'-') from dual;
column file_name format a50
column tablespace_name format a30
select file_id, file_name, tablespace_name, bytes/1024/1024 as "size(m)", status, autoextensible from dba_temp_files;
```

临时文件 id,临时文件名称和路径,表空间名,临时文件大小、状态、是否自动扩展。

(18) 统计哪些数据文件读写最频繁

```
select rpad('the count of read/write',100,'-') from dual;
col name for a40
select df.name,phyrds,phywrts
from v$filestat fs,v$datafile df
where fs.file# = df.file#
order by phyrds,phywrts;
NAME                                      PHYRDS    PHYWRTS
----------------------------------------  --------  --------
/oracle/oradata/yingshudb/users01.dbf         1         0
/oracle/oradata/yingshudb/fuzong01.dbf        1         0
/oracle/oradata/yingshudb/undotbs01.dbf      28        60
/oracle/oradata/yingshudb/sysaux01.dbf      618       219
/oracle/oradata/yingshudb/system01.dbf     3146        33
```

数据文件名称和路径、物理读总数、物理写总数。通过这个语句可以查出哪些数据文件读写最频繁,如果有 I/O 瓶颈,分散这些频繁读写的对象也是一个解决办法。

(19) 查询控制文件的名称

```
select * from v$controlfile;
STATUS     NAME                                         IS_RECOVE  BLOCK_SIZE  FILE_SIZE_BLKS
---------  -------------------------------------------  ---------  ----------  --------------
           /oracle/oradata/yingshudb/control01.ctl      NO             16384             594
           /oracle/oradata/yingshudb/control02.ctl      NO             16384             594
```

(20) 查询日志文件

```
set head on
set linesize 132
col status for a20
col member for a38
select vlf.member,vl.group#,vl.thread#,vl.bytes/1024/1024 as "size(m)",vl.status,vlf.
type from v$log vl,v$logfile vlf
where vl.group# = vlf.group#;

MEMBER                                    GROUP#    THREAD#    size(M)  STATUS      TYPE
----------------------------------------  --------  ---------  -------  ----------  ------
/oracle/oradata/yingshudb/redo01.log         1          1         500   INACTIVE    ONLINE
/oracle/oradata/yingshudb/redo02.log         2          1         500   CURRENT     ONLINE
/oracle/oradata/yingshudb/redo03.log         3          1         500   INACTIVE    ONLINE
/oracle/oradata/yingshudb/redo04.log         4          1         500   INACTIVE    ONLINE
```

6 列分别为:日志文件的位置和名称、一共有多少组、属于哪一个实例、日志文件大小、日志文件状态、是否在线。

(21) 表空间信息

```
select rpad('tablespace info',100,'-') from dual;
set head on
```

```
select tablespace_name tbs_name,block_size,extent_management
ext_mag,allocation_type,segment_space_management seg_space from dba_tablespaces;
TBS_NAME        BLOCK_SIZE    EXT_MAG    ALLOCATION_TYPE    SEG_SPACE
SYSTEM          8192          LOCAL      SYSTEM             MANUAL
SYSAUX          8192          LOCAL      SYSTEM             AUTO
UNDOTBS1        8192          LOCAL      SYSTEM             MANUAL
TEMP            8192          LOCAL      UNIFORM            MANUAL
USERS           8192          LOCAL      SYSTEM             AUTO
FUZONG          8192          LOCAL      SYSTEM             AUTO
6 rows selected.
```

5列分别为：表空间名称、块大小、表空间区管理方式、区分配方式、段空间管理方式。

（22）表空间剩余百分比

```
select rpad('tablespace use',100,'-') from dual;
set head on
column pct_free format a10
select tablespace_name, sum_m as "sum(m)" , sum_free as "sum_free(m)",to_char(100 * sum_
free/sum_m, '99.99') || '%' as pct_free
from ( select tablespace_name, sum(bytes)/1024/1024 as sum_m from dba_data_files group by
tablespace_name),
( select tablespace_name as ts_name, sum(bytes/1024/1024) as sum_free from dba_free_space
group by tablespace_name )
where tablespace_name = ts_name ( + )
union
select tablespace_name , sum_m as "sum(m)" , sum_free as "sum_free(m)" , to_char(100 * sum_
free/sum_m, '99.99') || '%' as pct_free
from (select tablespace_name, sum(bytes)/1024/1024 as sum_m from dba_temp_files group by
tablespace_name),
(select tablespace_name as ts_name, sum(bytes_free /1024/1024) as sum_free from v$temp_space
_header group by tablespace_name)
where tablespace_name = ts_name ( + )
order by pct_free;
TABLESPACE_NAME         sum(M)      sum_free(M)    PCT_FREE
--------------- -------     -----------    --------

SYSTEM                  700         427.625        61.09%
TEMP                    20          15             75.00%
USERS                   5           4              80.00%
SYSAUX                  600         486.75         81.13%
FUZONG                  10240       9247           90.30%
UNDOTBS1                5660        5647.625       99.78%

6 rows selected.
```

4列分别为：表空间名称、表空间总大小、剩余空间的大小、剩余百分之多少。

（23）表空间利用率

```
set pages 100
set linesize 200
```

```
col ts_name form a20 head 'tablespace'
col pieces form 99990 head 'pcs'
col ts_size form 99999,990 head 'sizeMB'
col largestpc form 999,990 head 'lrgMB'
col totalfree form 99999,990 head 'freeMB'
col pct_free form 990 head ' % free'
col whatsused form 999999,990 head 'used'
col pct_used form 990 head ' % used'
col problem head 'prob??'

select q2.other_tname ts_name, pieces, ts_size ts_size,
       nvl(largest_chunk,0) largestpc, nvl(total_free,0) totalfree,
       nvl(round((total_free/ts_size) * 100,2),0) pct_free,
       ts_size - total_free whatsused,
       nvl(100 - round((total_free/ts_size) * 100,2),100) pct_used,
       decode(nvl(100 - round((total_free/ts_size) * 100,0),100),
           85,' + ',86,' + ',87,' + ',88,' + ',89,'++',90,'++',91,'++',
           92,'++',93,'++',94,'++ + ',95,'++ + ',96,'++ + ',97,'++++',
           98,'++++ + ',99,'++++ + ',100,'++++ + ','') problem
from (select dfs.tablespace_name,count( * ) pieces,
             round(max(dfs.bytes)/1024/1024,2) largest_chunk,
             round(sum(dfs.bytes)/1024/1024,2) total_free
        from dba_free_space dfs group by tablespace_name) q1,
     (select tablespace_name other_tname,
             round(sum(ddf2.bytes)/1024/1024,2) ts_size
        from dba_data_files ddf2 group by tablespace_name) q2
where q2.other_tname = q1.tablespace_name( + )
order by nvl(100 - round((total_free/ts_size) * 100,0),100) desc
/

Tablespace          Pcs     SizeMb    LrgMB      FreeMb   % Free   Used   % Used Prob??
----------------    ----    -------   -------    -------- ------   -----  ------ ------
JISHIYU                          6         0          0       0            100  +++++
SYSTEM                2        700       427        428      61     272     39
USERS                 1          5         4          4      80       1     20
SYSAUX                2        600       485        485      81     115     19
FUZONG                3     10,240     3,968      9,247      90     993     10
UNDOTBS1              7      5,660     3,149      5,623      99      37      1

6 rows selected.
```

9列分别为：表空间名称、空闲空间分布在多少个片上、表空间总大小、最大的空闲区、空闲剩余多少兆字节，剩余百分比，使用了多少兆字节，使用率的百分比，是否利用率过高。

当超过90％时，建议增加表空间容量。

 表空间 JISHIYU 利用率达到 100%，建议马上扩展此表空间。

(24) 表空间碎片

```
select rpad('tablespace fragments info',100,'-') from dual;
select tablespace_name,count(*),max(bytes)/1024/1024 from dba_free_space group by
tablespace_name order by 2;
TABLESPACE_NAME    COUNT(*) MAX(BYTES)/10 24/1024
--------------- ---------- -----------------
SYSAUX                   1            486.625
USERS                    1                  4
SYSTEM                   2                427
FUZONG                   3               3968
UNDOTBS1                12               3149
```

3列分别为：表空间名称、表空间碎片的多少、表空间最大的一块区有多大。如果第二列值较高，最大的一块区又很小，那说明表空间碎片现象严重。

(25) 用户信息

```
col username for a15
col default_tablespace format a15;
col account_status format a20;
col temporary_tablespace format a15;
col profile format a20
select username,account_status,default_tablespace,temporary_tablespace,profile from dba_
users;

USERNAME       ACCOUNT_STATUS       DEFAULT_TABLESP      TEMPORARY_TABLE      PROFILE
------         -----------          ---------            -------              ---------
SYS            OPEN                 SYSTEM               TEMP                 DEFAULT
SYSTEM         OPEN                 SYSTEM               TEMP                 DEFAULT
YING           OPEN                 FUZONG               TEMP                 DEFAULT
OUTLN          EXPIRED & LOCKED     SYSTEM               TEMP                 DEFAULT
APPQOSSYS      EXPIRED & LOCKED     SYSAUX               TEMP                 DEFAULT
DBSNMP         EXPIRED & LOCKED     SYSAUX               TEMP                 DEFAULT
WMSYS          EXPIRED & LOCKED     SYSAUX               TEMP                 DEFAULT
DIP            EXPIRED & LOCKED     USERS                TEMP                 DEFAULT
ORACLE_OCM     EXPIRED & LOCKED     USERS                TEMP                 DEFAULT

9 rows selected.
```

5列分别为：用户名、账号状态、默认表空间、临时表空间、使用的profile。
如果应用用户默认表空间指向了system或sysaus表空间，则建议用户更正。应用用户默认的表空间，不应指向system表空间。

(26) 是否有用户误将对象放在system表空间中

```
col owner for a20
col segment_name for a15
col segment_type for a15
select owner,segment_name,segment_type,bytes/(1024*1024) msize from dba_segments
where tablespace_name = 'SYSTEM'
```

```
and owner not in ('ANONYMOUS','CTXSYS','DBSNMP','HR','MDSYS','ODM','ODM_MTR','OE','OLAPSYS',
'ORDPLUGINS','ORDSYS','OUTLN','PM','PUBLIC','QS','QS_ADM','QS_CB','QS_CBADM','QS_CS','QS_ES',
'QS_OS','QS_WS','RMAN','SCOTT','SH','SYS','SYSTEM','WKPROXY','WKSYS','WMSYS','XDB');
OWNER           SEGMENT_NAME      SEGMENT_TYPE        MSIZE
------------    ----------------  ---------------     --------
SHU             A                 TABLE               .0625
```

 用户 shu 将表 A 放在了表空间 system 中, 建议移至其他应用表空间。

在 system 表空间建应用对象, 不符合规划, 影响性能, 不利于空间管理。

(27) 配置文件对资源的限制情况

```
col profile for a18
col resource_name for a30
col limit for a20
select * from dba_profiles order by profile;
PROFILE         RESOURCE_NAME               RESOURCE_TYPE       LIMIT
-----------     -------------------------   ----------------    ---------
DEFAULT         COMPOSITE_LIMIT             KERNEL              UNLIMITED
DEFAULT         PASSWORD_GRACE_TIME         PASSWORD            7
DEFAULT         CPU_PER_SESSION             KERNEL              UNLIMITED
DEFAULT         CPU_PER_CALL                KERNEL              UNLIMITED
DEFAULT         LOGICAL_READS_PER_SESSION   KERNEL              UNLIMITED
DEFAULT         LOGICAL_READS_PER_CALL      KERNEL              UNLIMITED
DEFAULT         IDLE_TIME                   KERNEL              UNLIMITED
DEFAULT         CONNECT_TIME                KERNEL              UNLIMITED
DEFAULT         PRIVATE_SGA                 KERNEL              UNLIMITED
DEFAULT         FAILED_LOGIN_ATTEMPTS       PASSWORD            10
DEFAULT         PASSWORD_LIFE_TIME          PASSWORD            180
DEFAULT         PASSWORD_REUSE_TIME         PASSWORD            UNLIMITED
DEFAULT         PASSWORD_REUSE_MAX          PASSWORD            UNLIMITED
DEFAULT         PASSWORD_VERIFY_FUNCTION    PASSWORD            NULL
DEFAULT         PASSWORD_LOCK_TIME          PASSWORD            1
DEFAULT         SESSIONS_PER_USER           KERNEL              UNLIMITED

16 rows selected.
```

4 列分别为: profile 的名称、资源的名称、资源的类型、限制的情况。

对安全要求比较高的用户, 可以考虑使用 profile 来设置密码的过期等控制, 也可以控制用户使用系统资源。

(28) 查询将权限授给所有用户的对象

```
select * from dba_col_privs
where grantee = 'public'
and owner not in
('ANONYMOUS','CTXSYS','DBSNMP','HR','MDSYS','ODM','ODM_MTR','OE','OLAPSYS','ORDPLUGINS','ORDSYS',
'OUTLN','PM','PUBLIC','QS','QS_ADM','QS_CB','QS_CBADM','QS_CS','QS_ES','QS_OS','QS_WS','RMAN',
'SCOTT','SH','SYS','SYSTEM','WKPROXY','WKSYS','WMSYS','XDB');
```

```
select * from dba_tab_privs
where grantee = 'public'
and owner not in
('ANONYMOUS','CTXSYS','DBSNMP','HR','MDSYS','ODM','ODM_MTR','OE','OLAPSYS','ORDPLUGINS','ORDSYS',
'OUTLN','PM','PUBLIC','QS','QS_ADM','QS_CB','QS_CBADM','QS_CS','QS_ES','QS_OS','QS_WS','RMAN',
'SCOTT','SH','SYS','SYSTEM','WKPROXY','WKSYS','WMSYS','XDB');

select * from dba_role_privs
where grantee = 'PUBLIC';

select * from dba_sys_privs
where grantee = 'PUBLIC';
```

4个语句分别为：数据库列上的所有授权、数据库对象上的所有权限、显示已授予用户或其他角色的角色、已授予用户或角色的系统权限。

(29) 用户所拥有的角色

```
select rpad('the privilege of users',100,'-') from dual;
column username              format a15
column account_status        format a10
column default_tablespace    format a15
column temporary_tablespace  format a10
column granted_role          format a15
column privilege             format a23
set linesize 132
select du.username,du.account_status,du.default_tablespace,du.temporary_tablespace,drp.
granted_role,dsp.privilege
from dba_users du,dba_role_privs drp,dba_sys_privs dsp
where du.username = drp.grantee
and du.username = dsp.grantee
and du.username not in
('ANONYMOUS','CTXSYS','DBSNMP','HR','MDSYS','ODM','ODM_MTR','OE','OLAPSYS','ORDPLUGINS','ORDSYS',
'OUTLN','PM','PUBLIC','QS','QS_ADM','QS_CB','QS_CBADM','QS_CS','QS_ES','QS_OS','QS_WS','RMAN',
'SCOTT','SH','SYS','SYSTEM','WKPROXY','WKSYS','WMSYS','XDB');

USERNAME        ACCOUNT_ST  DEFAULT_TABLESP  TEMPORARY_  GRANTED_ROLE  PRIVILEGE
--------------- ----------- ---------------- ----------- ------------- --------------------
YING            OPEN        FUZONG           TEMP        RESOURCE      UNLIMITED TABLESPACE
YING            OPEN        FUZONG           TEMP        CONNECT       UNLIMITED TABLESPACE
YING            OPEN        FUZONG           TEMP        DBA           UNLIMITED TABLESPACE
... ...
9 rows selected.
```

6列分别为：用户名、账号状态、默认表空间、临时表空间、被授予的角色。

(30) 被授予 DBA 角色的用户

```
column username          format a15
column account_status    format a10
```

```
column default_tablespace      format a15
column temporary_tablespace format a10
column granted_role            format a15
column privilege               format a23
set linesize 132
select distinct
du.username,du.account_status,du.default_tablespace,du.temporary_tablespace,drp.granted_
role from dba_users du,dba_role_privs drp,dba_sys_privs dsp
where du.username = drp.grantee
and du.username = dsp.grantee
and drp.granted_role = 'DBA'
and du.username not in
('ANONYMOUS','CTXSYS','DBSNMP','HR','MDSYS','ODM','ODM_MTR','OE','OLAPSYS','ORDPLUGINS','ORDSYS',
'OUTLN','PM','PUBLIC','QS','QS_ADM','QS_CB','QS_CBADM','QS_CS','QS_ES','QS_OS','QS_WS','RMAN',
'SCOTT','SH','SYS','SYSTEM','WKPROXY','WKSYS','WMSYS','XDB');
USERNAME           ACCOUNT_ST DEFAULT_TABLESP TEMPORARY_ GRANTED_ROLE
--------------     ---------- --------------- ---------- ------------
TIEXIN             OPEN       JISHIYU         TEMP       DBA
SHU                OPEN       SYSTEM          TEMP       DBA
YING               OPEN       FUZONG          TEMP       DBA
```

6 列分别为：用户名、账号状态、默认表空间、临时表空间、被授予了 DBA 角色。

（31）默认表空间为 system 的用户

```
col username for a20
col default_tablespace for a30
col temporary_tablespace for a25
select username,default_tablespace,temporary_tablespace from dba_users
where username not in ('SYS','SYSTEM');
USERNAME              DEFAULT_TABLESPACE         TEMPORARY_TABLESPACE
---------------       -------------------------  --------------------
MGMT_VIEW             SYSTEM                     TEMP
DBSNMP                SYSAUX                     TEMP
SYSMAN                SYSAUX                     TEMP
SHU                   SYSTEM                     TEMP
```

 用户 shu 将默认表空间指向了 system，建议 shu 将默认表空间改为其应用规划建立的表空间。

（32）数据库 db_link

```
col owner for a15
col db_link for a25
col username for a15
col conn_string for a35
select owner,db_link,username,host conn_string,created created_time
from dba_db_links;
SQL> select owner,db_link,username,host conn_string,created created_time
```

```
        from dba_db_links;

OWNER           DB_LINK           USERNAME        CONN_STRING         CREATED_T
--------------- ----------------- --------------- ------------------- ---------
PUBLIC          SHUCFGLINK        ISMPCFG         SHU1                11-JAN-14
PUBLIC          SHU_LINK          ISMPCFG         SHU2                22-NOV-14
```

此数据库中存在两个 link,具参连接路径参见 SHU1、SHU2。

(33) 解读密码文件

```
col username for a20
select * from v$pwfile_users;
```

(1) 如果密码文件不存在或者名称错误,v$pwfile_users 将查不到任何数据。

(2) 添加 sysdba 权限用户,会记录到密码文件和 v$pwfile_users 中。

(3) 到回收 sysdba 等权限用户,密码文件记录依然存在,但是 v$pwfile_users 中无对应记录。

(4) 能否远程登录,依照 v$pwfile_users 所查结果为准。

(34) 已经安装的组件及版本号

```
set linesize 132
set pages 999
col comp_name for a35
col version for a20
col status for a20
select comp_name, version, status from dba_registry;
COMP_NAME                           VERSION              STATUS
----------------------------------- -------------------- --------------------
OWB                                 11.2.0.4.0           VALID
Oracle Application Express          3.2.1.00.12          VALID
Spatial                             11.2.0.4.0           VALID
Oracle Multimedia                   11.2.0.4.0           VALID
... ...
Oracle Database Java Packages       11.2.0.4.0           VALID
OLAP Analytic Workspace             11.2.0.4.0           VALID
Oracle OLAP API                     11.2.0.4.0           VALID

18 rows selected.
```

如果还要查数据库当前的 patch,可以查询以下视图:dba_server_registry、dba_registry_history、product_component_version、v$version,也可以通过 ORACLE_HOME 下的 Opatch、opatch lsinv 命令查询具体的补丁版本。

例如:

```
select version,bundle_series,comments from dba_registry_history;
```

```
VERSION           BUNDLE_SERIES         COMMENTS
---------------   -----------------     ------------------
11.2.0.4          PSU                   Patchset 11.2.0.2.4
11.2.0.4          PSU                   Patchset 11.2.0.2.4
```

(35) 查看 listener 的状态

```
SQL> host lsnrctl stat

LSNRCTL for Linux: Version 11.2.0.4.0 - Production on 07-JAN-2015 13:32:47

Copyright (c) 1991, 2013, Oracle.  All rights reserved.

Connecting to (DESCRIPTION = (ADDRESS = (PROTOCOL = TCP)(HOST = yingshu)(PORT = 1521)))
STATUS of the LISTENER
------------------------
Alias                     LISTENER
Version                   TNSLSNR for Linux: Version 11.2.0.4.0 - Production
Start Date                07-JAN-2015 13:25:18
Uptime                    0 days 0 hr. 7 min. 30 sec
Trace Level               off
Security                  ON: Local OS Authentication
SNMP                      OFF
Listener Parameter File   /u01/app/oracle/product/10.2.0/db_1/network/admin/listener.ora
Listener Log File         /u01/app/oracle/diag/tnslsnr/yingshu/listener/alert/log.xml
Listening Endpoints Summary...
  (DESCRIPTION = (ADDRESS = (PROTOCOL = tcp)(HOST = yingshu)(PORT = 1521)))
  (DESCRIPTION = (ADDRESS = (PROTOCOL = ipc)(KEY = EXTPROC1521)))
Services Summary...
Service "yingshu" has 1 instance(s).
  Instance "yingshu", status READY, has 1 handler(s) for this service...
Service "yingshuXDB" has 1 instance(s).
  Instance "yingshu", status READY, has 1 handler(s) for this service...
The command completed successfully
```

这里如果有服务名 yingshu,还有后面这个 handler(s),就可以证明我们的 listener 是好的,可以正常提供服务。我们还可以看一看后台进程在不在。

```
SQL> host ps -ef | grep tnslsnr
oracle    24087     1  0 13:25 ?        00:00:00 /u01/app/oracle/product/10.2.0/db_1/bin/tnslsnr LISTENER -inherit
oracle    24856 23885  0 13:39 pts/3    00:00:00 /bin/bash -c ps -ef | grep tnslsnr
oracle    24858 24856  0 13:39 pts/3    00:00:00 grep tnslsnr
```

(36) 收集 AWR 报告,生成问题 SQL 的执行计划

AWR(Automatic Workload Repository,自动工作负载库)的前身是 Oracle 9i 的 statspack,Oracle 8i 的时候叫 utlb、utle。发展到 Oracle 10g,推出了功能更加强大,信息更全面的 AWR。

```
SQL>@?/rdbms/admin/awrrpt
Current Instance
~~~~~~~~~~~~~~~~

  DB Id      DB Name      Inst Num Instance
--------- ------------ ------------------
2417604619 YINGSHU             1 yingshu


Specify the Report Type
~~~~~~~~~~~~~~~~~~~~~~~
Would you like an HTML report, or a plain text report?
Enter 'html' for an HTML report, or 'text' for plain text
Defaults to 'html'
Enter value for report_type: text -- 这里可以选择 text,方便在 Linux 下查看


Type Specified:                        text


Instances in this Workload Repository schema
~~~~~~~~~~~~~~~~~~~~~~~~~~~~~~~~~~~~~~~~~~~~

  DB Id     Inst Num DB Name      Instance      Host
---------- -------- ------------ ------------ --------
* 2417604619      1 YINGSHU       yingshu      yingshu
```

2.3 快速巡检脚本/方法

如果想让数据库巡检快速完成,例如去客户现场巡检,客户有十几套库,时间又给的很短。可以将以下脚本在 SQL PLUS 里运行,很快就可以生成巡检日志。

有了巡检日志,然后看一看警告日志里有没有 ORA-错误,操作系统的情况也要看一下。然后将日志记录下来,编写巡检报告。

脚本如下:

```
rem name
rem check.sql 2015-10-10
rem
rem description
rem collecting the db info
rem
rem notes
rem firstly connect system/xxxx or connect sys/xxx as sysdba
rem    secondly @check.sql
rem
rem    modified (yyyy-mm-dd)
```

```
## rem 是 remark 的简称,相当于 shell 脚本中的 #
prompt
prompt creating database report.
prompt this script must be run as a user with sysdba privileges.
prompt this process can take several minutes to complete.
prompt
set linesize 150
set timing off
set feedback off
alter session set nls_date_format = 'yyyy-mm-dd hh24:mi:ss';

rem spool report_keyinfo.txt

define filename = dbcheck

column dbname new_value _dbname noprint
select name dbname from v$database;

column spool_time new_value _spool_time noprint
select to_char(sysdate,'yyyymmdd') spool_time from dual;

pro
spool&FileName._&_dbname._&_spool_time..txt

set head off
select '以下是所有数据库相关的信息,生成的报告请用 UltraEdit 进行[修剪行尾空格]的操作'
from dual;
select 'DB INFO:' from dual;
set head on
select name as "db name",instance_name,open_mode,log_mode,status,host_name from v$database,
v$instance;
set head off
select 'the count of cpu:          ' || value from v$parameter where name = 'cpu_count';
select name from v$datafile where rownum < 2;
select 'db component version:      ' || product || ' ' || version from product_component_version
where upper(substr(product,1,3)) in ('ORA','TNS');

select 'sum of datafile size(g):   ' || trunc(sum(bytes)/1024/1024/1024) || 'g' as sum_g from
dba_data_files;



show sga;
select 'db_cache_size(m):          ' || value/1024/1024 from v$parameter where name = 'db_
cache_size';
select 'log_buffer(k):             ' || value/1024 from v$parameter where name = 'log_buffer';
select 'shared_pool_size(m):       ' || value/1024/1024 from v$parameter where name = 'shared
_pool_size';
```

```sql
select 'java_pool_size(m):           ' || value/1024/1024 from v$parameter where name = 'java_pool_size';
select 'large_pool_size(m):          ' || value/1024/1024 from v$parameter where name = 'large_pool_size';
select 'db_block_size(k):            ' || value/1024 from v$parameter where name = 'db_block_size';
select 'pga_aggregate_target(m): ' || value/1024/1024 from v$parameter where name = 'pga_aggregate_target';
select 'the count of tablespaces: ' || count(*) from v$tablespace;
select 'the count of datafiles:   ' || count(*) from v$datafile;
select 'the count of controlfiles:' || count(*) from v$controlfile;
select 'the size of redolog(m):   ' || bytes/1024/1024 from v$log;
set head on
select 'redo group:                  ',v1.* from v$log v1;
select 'redo member:                 ' from dual;
column member format a50
select * from v$logfile;
set head off

select 'yesterday redo average switch frequency(min) : ' || round(period/y.cnt,2) from
(select sum((a.first_time - b.first_time) * 24 * 60) as period
from v$log_history a, v$log_history b
where a.recid = b.recid + 1
and a.first_time > trunc(sysdate - 1) + 10/24
and a.first_time < trunc(sysdate - 1) + 16/24) x,
(select count(*) as cnt from v$log_history
where first_time > trunc(sysdate - 1) + 10/24
and first_time < trunc(sysdate - 1) + 16/24) y;

select rpad('arch info',100,'-') from dual;
select 'log_mode:                    ' || log_mode from v$database;
show parameter log_archive_dest

select rpad('concurrent user info',100,'-') from dual;
set head on
select sessions_max, sessions_warning, sessions_current, sessions_highwater, users_max from v$license;
set head off

select rpad('db components',100,'-') from dual;
set pagesize 500
column parameter format a40
column value format a30
set head on
select * from v$option;
set head off

select rpad('key parameters',100,'=') from dual;
column name format a50
column value format a50
select rpad('default parameters',50,'-') from dual;
```

```
set head on
select num,name,value from v$parameter where isdefault = 'TRUE';
set head off
select rpad('no default parameters',50,'-') from dual;
set head on
select num,name,value from v$parameter where isdefault = 'FALSE';
set head off

show sga;
select 'db_cache_size(m): ' || value/1024/1024 from v$parameter where name = 'db_cache_size';
select 'log_buffer(k): ' || value/1024 from v$parameter where name = 'log_buffer';
select 'shared_pool_size(m): ' || value/1024/1024 from v$parameter where name = 'shared_pool_size';
select 'java_pool_size(m): ' || value/1024/1024 from v$parameter where name = 'java_pool_size';
select 'large_pool_size(m): ' || value/1024/1024 from v$parameter where name = 'large_pool_size';

set head on
select pool,name,bytes/1024/1024 as "size(m)" from v$sgastat where name like '%free%';
set head off

select rpad('invalid objects',100,'-') from dual;
column owner format a15
column object_name format a25
column object_type format a15
column status format a10
select owner,object_name,object_type,status from dba_objects where status = 'INVALID';
select rpad('datafile is offline ? ',100,'-') from dual;
set head on
Select file_name from dba_data_files where status = 'OFFLINE'
set head off

select rpad('invalid constraints ',100,'-') from dual;
set head on
select owner, constraint_name, table_name, constraint_type, status from dba_constraints where status = 'DISABLED'
set head off

select rpad('invalid triggers ',100,'-') from dual;
set head on
select owner, trigger_name, table_name, status from dba_triggers where status = 'DISABLED'
set head off

select rpad('datafile info',100,'-') from dual;
column file_name format a50
column tablespace_name format a30
set head on
```

```
select file_id, file_name, tablespace_name, bytes/1024/1024 as "size(m)", status, autoextensible from dba_data_files;
set head off

select rpad('tmpefile info',100,'-') from dual;
set head on
select file_id, file_name, tablespace_name, bytes/1024/1024 as "size(m)", status, autoextensible from dba_temp_files;
set head off

select rpad('the count of read/write',100,'-') from dual;
set head on
select df.name,phyrds,phywrts
from v$filestat fs,v$datafile df
where fs.file# = df.file#
order by phyrds,phywrts;

select * from v$controlfile;
set head off

select rpad('REDO FILE',100,'-') from dual;
set head on
select vlf.member, vl.group#, vl.thread#, vl.bytes/1024/1024 as "size(m)", vl.status, vlf.type
from v$log vl,v$logfile vlf
where vl.group# = vlf.group#;
set head off

select rpad('tablespace info',100,'-') from dual;
set head on
select tablespace_name, block_size, extent_management, allocation_type, segment_space_management from dba_tablespaces;
set head off

select rpad('tablespace use',100,'-') from dual;
set head on
column pct_free format a10
select tablespace_name, sum_m as "sum(m)", sum_free as "sum_free(m)",to_char(100 * sum_free/sum_m, '99.99') || '%' as pct_free
from ( select tablespace_name, sum(bytes)/1024/1024 as sum_m from dba_data_files group by tablespace_name),
( select tablespace_name as ts_name, sum(bytes/1024/1024) as sum_free from dba_free_space group by tablespace_name )
where tablespace_name = ts_name (+)
union
select tablespace_name, sum_m as "sum(m)", sum_free as "sum_free(m)", to_char(100 * sum_free/sum_m, '99.99') || '%' as pct_free
from (select tablespace_name, sum(bytes)/1024/1024 as sum_m from dba_temp_files group by tablespace_name),
```

```sql
(select tablespace_name as ts_name, sum(bytes_free /1024/1024) as sum_free from v$temp_space
_header group by tablespace_name)
where tablespace_name = ts_name ( + )
order by pct_free;
set head off

select rpad('tablespace fragments info',100,'-') from dual;
set head on
select tablespace_name, count ( * ), max (bytes)/1024/1024 from dba_free_space group by
tablespace_name order by 2;
set head off

column default_tablespace format a15;
column account_status format a20;
column temporary_tablespace format a15;
select rpad('users info',100,'-') from dual;
set head on
select username, account_status, default_tablespace, temporary_tablespace, profile from dba_
users;
set head off

select rpad('the user objects in system tablespace',100,'-') from dual;
set head on
select owner, segment_name, segment_type, bytes/(1024*1024) msize from dba_segments
where tablespace_name = 'system'
and owner not in ('ANONYMOUS','CTXSYS','DBSNMP','HR','MDSYS','ODM','ODM_MTR','OE','OLAPSYS',
'ORDPLUGINS','ORDSYS','OUTLN','PM','PUBLIC','QS','QS_ADM','QS_CB','QS_CBADM','QS_CS','QS_ES',
'QS_OS','QS_WS','RMAN','SCOTT','SH','SYS','SYSTEM','WKPROXY','WKSYS','WMSYS','XDB');
set head off

select rpad('profile info',100,'-') from dual;
set head on
select * from dba_profiles order by profile;
set head off

select rpad('the roles and privs witch granted to public',100,'-') from dual;
set head on
select * from dba_col_privs
where grantee = 'public'
and owner not in
('ANONYMOUS','CTXSYS','DBSNMP','HR','MDSYS','ODM','ODM_MTR','OE','OLAPSYS','ORDPLUGINS',
'ORDSYS','OUTLN','PM','PUBLIC','QS','QS_ADM','QS_CB','QS_CBADM','QS_CS','QS_ES','QS_OS','QS_
WS','RMAN','SCOTT','SH','SYS','SYSTEM','WKPROXY','WKSYS','WMSYS','XDB');
select * from dba_tab_privs
where grantee = 'PUBLIC'
and owner not in
('ANONYMOUS','CTXSYS','DBSNMP','HR','MDSYS','ODM','ODM_MTR','OE','OLAPSYS','ORDPLUGINS',
'ORDSYS','OUTLN','PM','PUBLIC','QS','QS_ADM','QS_CB','QS_CBADM','QS_CS','QS_ES','QS_OS','QS_
WS','RMAN','SCOTT','SH','SYS','SYSTEM','WKPROXY','WKSYS','WMSYS','XDB');
select * from dba_role_privs
```

```sql
where grantee = 'PUBLIC';
select * from dba_sys_privs
where grantee = 'PUBLIC';
set head off

select rpad('the privilege of users',100,'-') from dual;
column username format a15
column account_status format a10
column default_tablespace format a15
column temporary_tablespace format a15
column granted_role format a30
column privilege format a30
set head on
select du.username, du.account_status, du.default_tablespace, du.temporary_tablespace, drp.
granted_role, dsp.privilege
from dba_users du, dba_role_privs drp, dba_sys_privs dsp
where du.username = drp.grantee
and du.username = dsp.grantee
and du.username not in
('ANONYMOUS','CTXSYS','DBSNMP','HR','MDSYS','ODM','ODM_MTR','OE','OLAPSYS','ORDPLUGINS',
'ORDSYS','OUTLN','PM','PUBLIC','QS','QS_ADM','QS_CB','QS_CBADM','QS_CS','QS_ES','QS_OS','QS_
WS','RMAN','SCOTT','SH','SYS','SYSTEM','WKPROXY','WKSYS','WMSYS','XDB');

set head off
select rpad('the users with dba role',100,'-') from dual;
set head on
select distinct du.username, du.account_status, du.default_tablespace, du.temporary_
tablespace, drp.granted_role
from dba_users du, dba_role_privs drp, dba_sys_privs dsp
where du.username = drp.grantee
and du.username = dsp.grantee
and drp.granted_role = 'DBA'
and du.username not in
('ANONYMOUS','CTXSYS','DBSNMP','HR','MDSYS','ODM','ODM_MTR','OE','OLAPSYS','ORDPLUGINS',
'ORDSYS','OUTLN','PM','PUBLIC','QS','QS_ADM','QS_CB','QS_CBADM','QS_CS','QS_ES','QS_OS','QS_
WS','RMAN','SCOTT','SH','SYS','SYSTEM','WKPROXY','WKSYS','WMSYS','XDB');
set head off

select rpad('db cache hit ratio',100,'-') from dual;
select 100-(a.value-b.value-c.value)/(d.value+e.value-b.value-c.value)
from
v$sysstat a,
v$sysstat b,
v$sysstat c,
v$sysstat d,
v$sysstat e
where a.name = 'physical reads'
and b.name = 'physical reads direct'
and c.name = 'physical reads direct (lob)'
and d.name = 'consistent gets'
```

```sql
and e.name = 'db block gets';

select 'sort_area_size hit ratio' from dual;
select a.value/(a.value + b.value)
from v$sysstat a, v$sysstat b
where a.name = 'sorts (disk)'
and b.name = 'sorts (memory)';

select 'log_buffer hit ratio' from dual;
select a.value/b.value
from v$sysstat a, v$sysstat b
where a.name = 'redo buffer allocation retries'
and b.name = 'redo entries';

set head on
select namespace,gethitratio,pinhitratio,reloads
from v$librarycache
where namespace in ('SQL AREA','TABLE/PROCEDURE','BODY','TRIGGER');

select sum(reloads)/sum(pins)
from v$librarycache;

select 1 - (sum(getmisses)/sum(gets))
from v$rowcache;
set head off

column p1text format a20
column p2text format a20
column p3text format a20
column event format a30
select rpad('wait events portion',100,'-') from dual;
set head on
select * from v$session_wait where event not like '%SQL*Net%';

show parameter back;

select value from v$parameter where name = 'background_dump_dest';

select name from v$database;

select name
from (select p2 latch# from v$session_wait where event in ('latch free')) b,v$latchname a
where a.latch# = b.latch#;

select addr,latch#,name,gets,spin_gets from v$latch order by spin_gets;

define filename = dbcheck

column dbname new_value _dbname noprint
select name dbname from v$database;
```

```
column spool_time new_value _spool_time noprint
select to_char(sysdate,'yyyymmdd') spool_time from dual;

spool off

rem set markup html off

rem set termout on

rem prompt
rem prompt output written to: &filename._&dbname._&_spool_time..txt

rem exit;
```

第 3 章 数据库系统的规划

3.1 概述

系统的成败取决于是否有优秀的规划和实施。我们决定要使用 Oracle 数据库,接下来就要规划这个服务系统。笔者建议要从成熟度和稳定性上重点考虑,保证服务系统能够按照客户的计划和要求提供服务,即使要求 7×24 小时的服务能力,系统也能轻松应对。

操作系统也很重要,操作系统软件控制和协调计算机及外部设备,支持应用软件开发和运行的系统,主要功能是调度、监控和维护计算机系统;负责管理计算机系统中各种独立的硬件,使得它们可以协调工作。系统软件使得计算机使用者和其他软件将计算机当作一个整体而不需要顾及到底层每个硬件是如何工作的。系统软件的选型也要考虑可扩展性,如果将来资源不够用,我们可以横向加硬件。

总之将速度充分地体现在桌面上,让 OLTP 终端的使用者,不用等待界面的弹出,让 OLAP 的使用者尽快拿到结果报表。

3.2 规划项目时的考虑因素

1. 充足的资源

稳健的服务系统,资源充足是关键,不能跑着跑着没内存了,CPU 满了。但现在我们新买的硬件一般都配置较高,老的硬件跑新的软件有时候就资源不够了。CPU 利用率高峰时期最好别超过 70%。各 CPU 之间利用率应该负载均衡。

Oracle 是很占内存的,Oracle 11g RAC 比 Oracle 10g 更占内存,有时候我们为了让速度更快,需要把整张表放在内存里面,减少 I/O 压力或根除 I/O 瓶颈。这时候更需要充足的内存资源。一般来说,内存大于 8GB,我们就建议使用 Linux 或 UNIX 操作系统了。有的用户买了 64GB 的内存,我们巡检的时候发现实际上才用了 4GB,浪费了资源,充足的资源也没体现到速度上去。也有的就 8GB 内存装 Oracle 11g RAC,CPU 内存经常满,参数已经设置得很合理了,但资源不够用,只能迁走一部分应用,把压力降下来。具体用多大的内存多少个 CPU 还是取决于系统的繁忙程度、并发数,这方面提前要计算好。

2. 存储的考虑

存储的方式可以采用较成熟的文件系统——ASM。ASM 是 Oracle 10g 推出的技术,Oracle 10g 的时候可以用裸设备,也可以用 ASM。Oracle 10g 的 vote 和 OCR 盘不能放在 ASM 上,所以如果还用 Oracle 10g ASM,就得单独做 2 个或 6 个裸设备给 vote 和 OCR 用。Oracle 11g 就完全解决了这个问题,况且 Oracle 11g R2 抛弃了裸设备,只能用 ASM 或文件系统。当然了数据库的 redo 的日志最好不要放在 RAID5 或 RAID6 上,最好放在 RAID1 或 RAID10 上,提高频繁的写速度。存储级别划的 lun 给 ASM 用的时候,大小不能大于 2TB,大于 2TB 的时候,Oracle 软件容易出问题。一般我们买的物理盘也都小于 2TB。在存储级别和数据库级别结合起来考虑 I/O 的分散。

生产库磁盘最好采用 SAS 2.0 接口,转速 15000 转/分的作为存储硬盘,充分保证 I/O 速度。以前在做调优项目的时候,有的瓶颈就是在 I/O 上,换高速盘提高 I/O 速度是一个解决思路,将常用的表和索引 cache 到内存里减少物理 I/O 也是一个解决思路。

存储使用两个存储,一个生产存储,一个备份存储。备份存储要求容量大,速度要求没有生产存储那么苛刻,可以采用 SATA 盘。

3. 软件的版本

安装较成熟稳定的数据库版本,升级应用最新的 patch。操作系统补丁要打齐,数据库补丁也一定要打全,最好补丁升级到最新,以保障系统的稳定性,也减少了再次升级补丁给系统带来的隐患。上次在联通做升级时,升级后发现速度提高了很多,看来升级不仅治 bug,还能提高速度。数据库版本当然是在一开始在从零开始的时候就安装到最后的一个补丁,一开始就做得很好,不要等到出事了再各种处理各种升级,亡羊补牢成本很高,风险也大。

数据库的版本这里也列一下,我们要安装这样的版本,数据库才能健康不出事、少出事。

(1) Oracle 8i 要安装至 8.1.7.4。

(2) Oracle 9i 要安装到 9.2.0.8.7。

(3) Oracle 10g 要升级至 10.2.0.5.12。

(4) Oracle 11g 安装 11.2.0.4,然后打补丁至 11.2.0.4.6。

(5) Oracle 12c 安装 12.1.0.2.3,Windows 是 12.1.0.2.4。

Oracle 11g 的补丁还要更新最新的补丁,在安装的时候到 https://support.oracle.com 上下载。当然下载补丁要有一个账号和密码,Oracle 也是通过控制补丁来控制正版化。Oracle 11g 现在已经是很稳定的版本,目前新上的系统主要都是 Oracle 11g。

Oracle 12c 因为是最新的版本,目前生产的还比较少,可能 bug 很多。如果思想比较超前上了 Oracle 12c,遇到新的问题,要提交给 Oracle 去处理,成为 bug 的践行者。

实施的人员,一定要用这方面的技术专家,据我观察,大部分的安装实施由于工程师水平或责任心问题是不给客户安装小补丁的,甚至就装一个 10.2.0.1 就应付了事。甲方在项目规划期间也要强调版本补丁这个事儿。所以要考核实施人员的技术水平,杜绝选择了好的软件,而因为实施技术水平导致信息系统问题的发生。况且在实施的过程中,就必须完成系统调优的部分工作。把各个地方都设计得非常优化,到后期运维的时候,调优就省去或减少这些问题了。所以在设计规划期,就有调优的内容了。

4. 操作系统

在计算机软件中最基本的就是操作系统(Operating System,OS)。它是最底层的软件,它控制所有计算机运行的程序并管理整个计算机的资源,是计算机裸机与应用程序及用户之间的桥梁。没有它,用户也就无法使用某种软件或程序。操作系统是计算机系统的控制和管理中心,从资源角度来看,它具有处理机、存储器管理、设备管理、文件管理等功能。

刚才说过,如果内存大于 8GB,建议采用 Linux 或 UNIX 操作系统,原因是:对于 Oracle 数据库来说,Windows 为单线程服务模式,而 Linux 为多进程服务模式。进程比线程速度要快,进程比线程可控制性强。用户可以独享一个进程,一个用户进程就有一个操作系统进程对它专门服务,而线程我们就无法控制。进程比线程内存使利用率高,我们可以观察一下内存的利用率,有时候你内存再大,Windows 似乎用不上。再者说,银行业电信业用的全是 Unix 至少是 Linux 做关键服务。Windows 肯定有它的局限之处。有的用户买了 96GB 内存,安装一个 Windows 操作系统,很难发挥它内存大的效能。Windows 实际占用内存只有 8GB。Windows 上跑 Oracle 数据库出了故障有时候也不好定位处理。服务器主流操作系统还是 Linux、UNIX。

从稳定性上分析,UNIX 也更胜一筹,巡检的时候,运行 5 年没有重启的 UNIX 都很常见。运行 5 年不重启的 Windows 确实不常见。稳定性 UNIX 强于 Linux,Linux 强于 Windows。

5. Oracle RAC 的节点数

Oracle RAC 理论上可以支持很多节点,Oracle 10g 第二版可以支持 100 个节点,Oracle 11g 也一样。讨论理论上的最多可以支持的节点数已经没有太大的意义。实际上我们认为节点越多出故障的几率越大,管理维护成本越高。RAC 用两个节点的用户最多,也有的用户使用 4 个节点或 6 个节点,甚至 8 个节点。再多的话,除了原厂,就没人建议用了,肯定不太好用。

RAC 本身又不能防止存储损坏,只能防止主机损坏。如果考虑横向扩展的话,负载均衡也是有条件的,不是随便拿一个应用连到 RAC 里面去,就可以实现负载均衡,负载均衡的条件是分开应用,几个节点各跑各的应用,各跑各的表,避免或减少交叉访问。如果就一个应用,必须交叉访问,那也行,但是你的压力不能太大,太大了就可能出现 gc 开头的等待事件,就又要做相应的调整。话又说回来,既然压力又不能太大,应用又很难分开,那做更多的节点就更没有必要了。所以说,RAC 肯定不是节点越多越好。RAC 的主要功能一是防止主机坏,二是解决 CPU 不够用的问题。

如果单机够用就用单机,然后做一个物理 DataGuard,既防主机坏又防存储坏,还可以应急容灾,Oracle 11g DataGuard 备库还可以当查询库。如果单机 CPU 不够用就做 2 个节点的 RAC,尽量把应用分开,减少心跳线网络的压力。如果用户财力足,技术人员多,水平高,应用也多,又想集中管理,那可以做成 4 个或 6 个节点,节点多了也是对运维人员的一个锻炼与考验。

6. 备份和容灾

数据库备份是必需的,不备份一旦出现问题,我们将无法或很难找回原来的数据,对企业的影响重大,损失可谓惨重,甚至是致命的。就算是我们个人的笔记本或台式机电脑,我也买两个 2TB 的大硬盘,养成定期备份的习惯,万一磁盘坏了或笔记本坏了,我可以快速地再买一个电脑,将大硬盘里的关键数据拷回新买的电脑里,很快地就可以恢复工作,我多年积累的工作经验记录、方案、脚本、视频都还在,心里踏实。

企业级的数据就更重要了,有可能重要到关乎企业生死,不但要备份,还要容灾,就算地震来了,火灾来了,洪水来了也不能毁坏我们的数据。我们要根据数据的重要程度,选择不同的备份和容灾方式。

带用的备份方式:❶RMAN 备份,可以不用停机,可以恢复至数据库损坏的那一刻,做到一条数据也不丢,是我们最常用的备份恢复方式。❷exp、imp 或 expdp、impdp 逻辑备份,特点是操作简单,只能恢复至开始备份的那一刻,对于备份之后变化的数据无能为力。常常作为数据迁移的方法。❸冷备,方法是关闭数据库,将数据文件、控制文件、redo 日志、spfile、pfile、密码文件复制至另一个地方,恢复的方法是再复制回来,也可以作为数据库迁移的方法,实际备份中不常用。❹利用备份软件,使用 nbu、commvault 等第三方软件调用 RMAN 命令,实现备份,第三方软件不仅可以备份数据库,还可以备份操作系统、应用程序,介于综合备份的考虑,有的企业就选择备份软件了。

制定备份策略,主要考虑下面几个方面:❶数据的重要程度;❷数据能不能再造;❸数据每天或每周的变化量有多大;❹数据的总量有多大;❺其他简单备份方式够不够用。以上几个问题搞清楚了,我们就知道了数据需要什么级别的备份容灾。省钱的备份办法就是 RMAN 定时做备份,其他重要可再造的程序定期手工复制走。操作系统、应用软件、数据库软件都可以再造,重新安装就行了,不用备份也没关系,我们只备份核心的重要数据。对于变化量不大的应用程序,不做实时备份。

3.3 项目的规划

下面我们就一个 2 节点 RAC 数据库系统做一个规划。

1. 硬件拓扑图(见图 3-1)

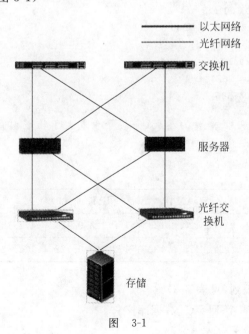

图 3-1

2. 操作系统安装规划（见表3-1）

表 3-1

业务系统名称	数据库服务器
操作系统或软件	Linux
操作类型	Redhat Linux Enterprise Server 5.8
内核版本	2.6.18-308.el5
系统工具及服务需求	应用服务器中间件 Weblogic 10
系统补丁包需求	参见官方安装文档
JDK	Java Version "1.6.0_22"
程序安装目录、数据空间目录	40GB
CPU 个数	4 颗 8 核
内存大小	64GB
数据库安装目录挂载点	/u01　大小为 100GB
启动目录	/boot/　大小为 1GB
Swap 分区	32GB
备份存储挂载	/backup　大小为 6TB

操作系统核心参数规划：

fs.aio-max-nr = 1048576
fs.file-max = 6815744
kernel.shmall = 2097152
kernel.shmmax = 34359738368
kernel.shmmni = 4096
kernel.sem = 250 32000 100 128
net.ipv4.ip_local_port_range = 9000 65500
net.core.rmem_default = 262144
net.core.rmem_max = 4194304
net.core.wmem_default = 262144
net.core.wmem_max = 1048586

说明：kernel.shmall 是很重要的参数，一般等于或大于物理内存的一半。
IP 规划如表 3-2 所示。

表 3-2

主　　机	主机名	SCAN IP	IP 地址	VIP	Private IP
数据库服务器 1	YS_DB_1	172.128.0.123	172.128.0.121	172.128.0.123	192.168.0.1
数据库服务器 2	YS_DB_2	172.128.0.123	172.128.0.122	172.128.0.124	192.168.0.2

3. 数据库存储系统规划（见表3-3）

采用单逻辑盘 100GB，ASM 管理数据库文件。

表 3-3

lun	用途	lun 大小	RAID 设置	设备类型
YS-lun01-01	rote, OCR	10GB	RAID6	ASM
YS-lun01-02	rote, OCR	10GB	RAID6	ASM
YS-lun01-03	rote, OCR	10GB	RAID6	ASM
YS-lun01-04	redo log	1000GB	RAID1	ASM
YS-lun01-05	data	1000GB	RAID6	ASM
YS-lun01-06	data	1000GB	RAID6	ASM
YS-lun01-07	data	1000GB	RAID6	ASM
YS-lun01-08	data	1000GB	RAID6	ASM
YS-lun01-09	index	1000GB	RAID6	ASM
YS-lun01-10	index	1000GB	RAID6	ASM
YS-lun01-11	index	1000GB	RAID6	ASM
YS-lun01-12	备用	1000GB	RAID6	ASM
YS-lun01-13	备用	1000GB	RAID6	ASM

4. 数据库安装规划（见表 3-4）

表 3-4

业务系统名称	核 心 生 产
Oracle 数据库版本	Oracle 11g 企业版, 11.2.0.4.6
PSU	p20299013 (11.2.0.4.6)
GI PSU	p20485808 (11.2.0.4.6)
SPU(CPU)	p20299015_112040_Linux-x86-64.zip
Oracle 安装基本目录 ORACLE_BASE	/u01/app
数据库安装目录 ORACLE_HOME	/u01/app/oracle/11.2.0.4
数据库名 ORACLE_SID	ysdb
实例名 1	ysdb1 核心生产应用使用
实例名 2	ysdb2 人力资源应用使用, 财务应用使用
数据库字符集	ZHS16GBK
Process 设置	1500
flash recovery area 回闪区	不使用回闪区
enterprise manager 企业管理器	不使用 IE 企业管理器 em
数据文件自动扩展功能	关闭数据文件自动扩展功能, 手工管理数据文件大小
数据存储格式, 归档日志位置	dgdata, dgarch
db_block_size	8KB
内存管理方式	关闭自动内存管理功能, 手工管理内存大小
统计信息收集	关闭, 定期手工收集统计信息
不安装数据库选件	Oracle JVM, Oracle XML DB, Oracle Multimedia,
不安装数据库选件	Oracle Application Eepress, Oracle Text, Oracle OLAP
不安装数据库选件	Oracle Spatial, Enterprise Manager Repository
Redo 日志组数大小	每个 thread 4 组, 合计 8 组, 每个组员 400MB
数据文件大小	system、sysaux、user 表空间, 每个数据文件 4GB
数据文件大小	temp、undo, 和其他所有数据文件每个 16GB
归档模式	打开, 归档日志放在 ASM 中的 arch 磁盘组

重要数据库参数设置如表 3-5 所示。

表 3-5

参　　数	参 数 值	参　　数	参 数 值
db_cache_size	26GB	sga_target	0
shared_pool_size	4GB	session_cached_cursors	200
java_pool_size	40MB	processes	1500
large_pool_size	200MB	undo_management	AUTO
log_buffer	60MB	undo_retention	18000
pga_aggregate_target	4GB	db_writer_processes	16
sga_max_size	40GB	open_cursors	1000

具体的安装步骤请参照官方文档：

◆ Oracle® Grid Infrastructure；
◆ Installation Guide；
◆ 11g Release 2 (11.2) for Linux；
◆ E17212-08。

以及：

◆ Oracle® Database；
◆ Installation Guide；
◆ 11g Release 2 (11.2) for Linux；
◆ E16763-06。

DBA 要会使用官方文档，快速地查资料、查命令。

5．数据库备份策略（见表 3-6）

在归档模式下，数据库采取每天全备的方式。

表 3-6

备份路径	此挂载点需要 200GB 空间，并且不和其他数据文件同在一个 vg。 /backup/ysdb/rmanbak

备份策略，每周全备。
backup_dbfull.rcv
	将以下脚本加到定时任务中，实现每天全备。
	run {
	backup database include current controlfile；
	backup archivelog all delete all input；
	}

RMAN 备份的参数设置	configure backup optimization on； configure channel device type disk format '/backup/ysdb/rmanbak/db_%u'； configure controlfile autobackup on； configure controlfile autobackup format for device type disk to '/backup/ysdb/rmanbak/cf_%f'； configure retention policy to recovery window of 21 days； configure default device type to disk； configure device type disk backup type to compressed backupset；

续表

定时任务设置	crontab - l 每天凌晨1点钟全备。 0 1 * * * sh /home/oracle/backupfull.sh
Backupfull.sh 脚本	set NLS_LANG=american ＞/dev/null 2＞&1 export ORACLE_SID=ysdb1 ＞/dev/null 2＞&1 rman target / nocatalog @ backup_dbfull.rcv log/backup/ysdb/rmanbak/backup_dbfull_`date -u ＋％Y％m％d`.log ＞/dev/null 2＞&1

第 4 章 安装RAC数据库

4.1 安装 RAC 前的准备工作

系统规划好之后,进入实施安装阶段。安装数据库之前,有很多准备工作需要按照规划做好,这样在安装的过程中才会一路绿灯。前期工作如果准备得不充分,一是安装过程会遇到这样那样的错误,安装过程就报错;二是安装的数据库有问题,有可能出现性能差、bug 多等隐患。安装前必须做好准备工作。

本章的实际操作环境是在 linux 上安装两节点 Oracle RAC。

参照第 3 章的规划,我们需要做以下准备工作。

(1) 硬件资源满足需求。例如要求最少 2.5GB 内存,一般要求大于 16GB,现在内存配置几百 GB 已经很常见了,Oracle 数据库对内存要求很高,内存大可以换取性能好。

(2) 存储按照大小划好 lun,映射到对应的主机上,实现多路径功能,在主机端能正常读写存储映射过来的磁盘。

(3) 网络环境配置并连接完毕。交换机、路由器按要求走好线,IP 地址设好 ping 通。

(4) 操作系统为 Oracle 认可的操作系统版本,Oracle 要求的操作系统包安装完成。

(5) 安装目录的文件系统建好,最好单独划一个挂载点给 Oracle 用,如叫/u01 或/oracle。

(6) 操作系统核心参数提前算好,改正。算法参见第 7 章。

(7) 创建用户 grid、oracle,创建组 oinstall、dba、asmoper、asmadmin 等。

(8) 创建目录,修改目录权限。

(9) 数据库规划,包括数据库名,是否归档,安装哪些组件,如何备份,是否 ASM 等。

(10) 安装过程请参考官方文档,官方文档可以在 oracle.com 上下载,文档是纯英文的。

◆ Oracle® Grid Infrastructure

◆ Installation Guide 11g Release 2 (11.2) for Linux E17212-08

以及：
- Oracle® Database
- Installation Guide 11g Release 2（11.2）for Linux E16763-06

4.2 删除 Linux 下的 Oracle 数据库

安装失败，旧库淘汰，产品升级之前等情况，都可能面临删除现有的 Oracle 数据库。关于删除 Oracle 有官方文档，Oracle 11g 还有一个专门用于删除的软件包，是我们下载补丁时的最后一个包。使用软件包删除的方法我们这里不做介绍了，我们总结一下手工删除 RAC 的过程。单机版的数据库也一样，基本上包含在以下的步骤中。删除 Oracle 10g 和删除 Oracle 11g 有些不一样，我们的实验是 Oracle 11g，Oracle 10g 的删除方法请参照官方文档 *How to Clean Up After a Failed Oracle Clusterware（CRS）Installation*。以下步骤，除了处理共享存储部分，其余 RAC 的每个节点都要操作。

（1）停止数据库，停止 crs：

```
SQL> shutdown abort;
ORACLE instance shut down.

# crsctl stop crs                        //使用 root 用户，RAC 每个节点执行
```

（2）删除文件，删除 ohasd，删除 /etc 下 ora 开头的文件：

```
[root@yingshu1 etc]# cd /etc/init.d
[root@yingshu init.d]# rm -f ohasd
[root@yingshu init.d]# rm -f init.ohasd
[root@yingshu1 etc]# cd /etc/
[root@yingshu1 etc]# rm -rf ora*
```

（3）检查是否还有 Oracle 进程存在，如果有使用 kill -9 删除：

```
ps -ef|grep ora_                         //查看是否有数据库进程
ps -ef|grep d.bin                        //查看 crs 进程
ps -ef|grep LOCAL=NO                     //查看远程连接到数据库进程
ps -ef|grep asm_                         //查看 ASM 进程
ps -ef|grep lsnr                         //查看 listener 进程
ps -ef|grep crs
ps -ef|grep evm
ps -ef|grep css
```

（4）如果没有其他 Oracle 软件在运行，则删除：

```
[root@yingshu1 tmp]# rm -rf /var/tmp/.oracle
[root@yingshu1 tmp]# rm -rf /tmp/.oracle
```

（5）确认没有其他软件在用 ORACLE_BASE。删除 grid 用户 ORACLE_BASE、ORACLE_HOME 下所有文件：

```
[root@yingshu app]# rm -rf /u01/app/*
```

(6) 删除 Oracle 用户 ORACLE_BASE、ORACLE_HOME 下所有文件：

```
[root@yingshu oracle]# rm -rf /oracle/app/*
```

(7) dd Oracle 曾经用过的磁盘，包括 ocr、votingdisk、asmdisk，如果你的 OCR、votingdisk 在共享文件系统上，那么手工删除：

```
dd if=/dev/zero of=/dev/raw/raw1 bs=1M count=10
dd if=/dev/zero of=/dev/raw/raw2 bs=1M count=10
dd if=/dev/zero of=/dev/raw/raw3 bs=1M count=10
dd if=/dev/zero of=/dev/raw/raw4 bs=1M count=10
dd if=/dev/zero of=/dev/raw/raw5 bs=1M count=10
dd if=/dev/zero of=/dev/raw/raw6 bs=1M count=10
10+0 records in                                          //输出结果
10+0 records out                                         //输出结果
10485760 bytes (10 MB) copied, 0.379184 seconds, 27.7 MB/s  //输出结果
```

(8) 最好重启一下操作系统，aix 使用命令 shutdown -Fr。删除的工作就完成了。

```
[root@yingshu dev]# reboot
```

4.3 RAC 安装过程

（1）选择合适的操作系统版本。按照官方文档的要求，选择合适的操作系统版本。版本低了不行，高了也不好，之后安装的时候会遇到因为版本问题的错误，表 4-1 给出了 Linux x86-64 操作的内核要求。

表 4-1

Linux Distribution	Requirements
Asianux Distributions	• Asianux Server 3, Service Pack 2 (SP2)
Enterprise Linux Distributions	• Enterprise Linux 4 Update 7, kernel 2.6.9 or later • Enterprise Linux 5 Update 2, kernel 2.6.18 or later
Red Hat Enterprise Linux Distributions	• Red Hat Enterprise Linux 4 Update 7, kernel 2.6.9 or later • Red Hat Enterprise Linux 5 Update 2, kernel 2.6.18 or later
SUSE Enterprise Linux Distributions	• SUSE 10, kernel 2.6.16.21 or later • SUSE 11, kernel 2.6.27.19 or later

（2）确认操作系统版本（两节点操作）：

```
[root@yingshu1 etc]# cat /etc/redhat-release
Red Hat Enterprise Linux Server release 5.8 (Tikanga)
[root@yingshu1 etc]#
[root@yingshu1 etc]# cat /etc/issue
Red Hat Enterprise Linux Server release 5.8 (Tikanga)
Kernel \r on an \m
```

（3）确认物理内存大小（两节点操作）：

```
[root@yingshu1 etc]# grep MemTotal /proc/meminfo
MemTotal: 32922156 kB
```

（4）确认 swap 空间大小，官方建议：如果内存在 2GB 以内，则等于内存的 1.5 倍；大于 2GB 一般等于内存大小；大于 16GB，可以设 swap 为 16GB。交换区等于或大于内存大小。我们这里设为 32GB。（两节点操作）

```
[root@yingshu1 etc]# grep SwapTotal /proc/meminfo
SwapTotal: 32922156 kB
```

（5）临时挂载点的大小，至少 1GB。（每个节点操作）

```
df -kh /tmp
Filesystem              Size  Used  Avail  Use%  Mounted on
/dev/mapper/VolGroup00-LogVol02
                        12G   15M   12G    0%    /tmp
```

（6）划分好存储，使用多路径软件，做好聚合。在存储中划分三个 ocr 盘，每个磁盘 5GB。其余全为数据盘，每个磁盘 1TB，或根据数据库的实际大小安排磁盘大小，如每个磁盘 500GB。不可以直接使用/dev/sdb，使用 sdb 这种直接认到的磁盘安装会失败，第二节点执行 root.sh 失败。我们使用/dev/raw/raw1, raw2……前三个是 ocr 盘，每个 10GB。从 raw4 开始，每个 1TB。

```
[root@yingshu1 init.d]# cd /dev/raw/
[root@yingshu1 raw]# ls -l
total 0
crw-rw---- 1 grid asmadmin 162, 1 Sep 4 11:13 raw1
crw-rw---- 1 grid asmadmin 162, 2 Sep 4 11:13 raw2
crw-rw---- 1 grid asmadmin 162, 3 Aug 31 09:04 raw3
crw-rw---- 1 grid asmadmin 162, 4 Aug 31 08:59 raw4
crw-rw---- 1 grid asmadmin 162, 5 Aug 31 08:59 raw5
```

（7）是否是 64 位操作系统。因为 32 位操作系统对内存使用有限制，所以服务器都不用 32 位操作系统，现在我们的笔记本大都是 64 位了，所以 32 位系统做服务器肯定不好。（每个节点操作）

```
#getconf LONG_BIT
64
```

（8）检查主机名，网络。（每个节点操作）

```
[root@yingshu1 etc]# hostname
yingshu1
# ifconfig -a
# ping
```

（9）配置并检查/etc/hosts 配置。我们把 scan IP 固定写在 hosts 文件里。（每个节点操作）

注意 127.0.0.1 对应第二台的主机名是 yingshu2，其他两边一样。

```
# Do not remove the following line, or various programs
# that require network functionality will fail.
```

```
127.0.0.1           yingshu1 localhost.localdomain localhost
::1                 localhost6.localdomain6 localhost6

#public ip
192.168.1.10        yingshu1
192.168.1.11        yingshu2
#priv ip
10.10.10.211        yingshu1-priv
10.10.10.212        yingshu2-priv
#vip ip
192.168.1.13        yingshu1-vip
192.168.1.14        yingshu2-vip
#scan ip
192.168.1.12        yingshu-scan
```

（10）检查两个节点时钟是否同步，如果主机自己实现时钟同步服务 ntp，可以使用公司统一的时间服务器，不使用 Oracle 时钟同步服务。则 Oracle 忽略自己的时钟同步服务，不进行时钟同步，如果系统没有 ntp 服务，则 Oracle 启用活动的 CTSS 进程，以一台作为时钟同步服务器进行集群中的时间同步。Oracle 10g 和 Oracle 10g 之前，必须手工配置操作系统的时钟同步。我们这里使用 Oracle 自己的 ctss 服务，删除 ntp 服务。

停止 ntp 服务：

```
/etc/init.d/ntpd stop
chkconfig ntpd off
mv /etc/ntp.conf /etc/ntp.conf.org
```

删除 ntp 服务，根据提示，输入 yes，回车：

```
[root@yingshu1 etc]# date
Fri Jul 31 10:39:18 CST 2015
[root@yingshu1 ~]# yum remove ntp
```

（11）检查是否有 Oracle 进程存在，如果我们从一个非全新的系统安装，或以前安装失败过，机器上留有原来安装过的痕迹，一定要删除干净，删不干净下次肯定还装不上。（每个节点操作）

```
ps -ef|grep ora_              //查看是否有数据库进程
ps -ef|grep d.bin             //查看 crs 进程
ps -ef|grep LOCAL=NO          //查看远程连接到数据库进程
ps -ef|grep asm_              //查看 ASM 进程
ps -ef|grep lsnr              //查看 listener 进程
```

（12）检查操作系统是否已安装必要的补丁包，以下是在 Red Hat Enterprise Linux 5 上安装 11.2.0.4 Grid Infrastructure 11g 所需的操作系统软件包，如果是安装 11.2.0.1 还需要 32 位的软件包，不过一般没有人再装 11.2.0.1 了。使用 rpm -qa 检查是否有这个软件包。（每个节点操作）

```
rpm -qa | grep elfutils-libelf
rpm -qa | grep elfutils-libelf-devel
```

```
rpm -qa | grep gcc
rpm -qa | grep gcc-c++
rpm -qa | grep glibc
rpm -qa | grep glibc-common
rpm -qa | grep glibc-devel
rpm -qa | grep glibc-headers
rpm -qa | grep elfutils-libelf-devel-static
rpm -qa | grep kernel-headers
rpm -qa | grep binutils
rpm -qa | grep compat-libstdc++-296
rpm -qa | grep compat-libstdc++-33
rpm -qa | grep control-center
rpm -qa | grep ksh
rpm -qa | grep libaio
rpm -qa | grep libaio-devel
rpm -qa | grep libgcc
rpm -qa | grep libgnome
rpm -qa | grep libgnomeui
rpm -qa | grep libgomp
rpm -qa | grep libstdc++
rpm -qa | grep libstdc++-devel
rpm -qa | grep libXp
rpm -qa | grep make
rpm -qa | grep sysstat
rpm -qa | grep numactl
rpm -qa | grep unixODBC
rpm -qa | grep unixODBC-devel
```

(13) 如果缺失某个软件包,去安装光盘里/media/RHEL_5.8 x86_64 DVD/Server 这个文件夹下找。使用 rpm -ivh 安装,样例如下:(每个节点操作)

```
[root@yingshu1 tmp]# rpm -ivh unixODBC-devel-2.2.11-10.el5.x86_64.rpm
warning: unixODBC-devel-2.2.11-10.el5.x86_64.rpm: Header V3 DSA signature: NOKEY, key ID 37017186
Preparing...                ########################################### [100%]
   1:unixODBC-devel         ########################################### [100%]
```

(14) 建组,建用户。安装 RAC 的时候一定要把用户名和组的 ID 号(如 501)加上,否则两边 ID 号不一样会出问题。必须为 oracle、grid 用户设置密码,否则对等性检查不能通过。

```
[root@yingshu1 tmp]# groupadd -g 501 oinstall
[root@yingshu1 tmp]# groupadd -g 502 dba
[root@yingshu1 tmp]# groupadd -g 503 oper
[root@yingshu1 tmp]# groupadd -g 504 asmadmin
[root@yingshu1 tmp]# groupadd -g 505 asmoper
[root@yingshu1 tmp]# groupadd -g 506 asmdba
[root@yingshu1 tmp]# useradd -u 1101 -g oinstall -G dba,asmdba,oper oracle
[root@yingshu1 tmp]# passwd oracle
```

```
Changing password for user oracle.
New UNIX password:
Retype new UNIX password:
passwd: all authentication tokens updated successfully.
[root@yingshu1 tmp]# useradd -u 1102 -g oinstall -G asmadmin,asmdba,asmoper,oper,dba grid
[root@yingshu1 tmp]# passwd grid
Changing password for user grid.
New UNIX password:
Retype new UNIX password:
passwd: all authentication tokens updated successfully.
```

如果建立的用户有问题,那么删除的命令是 userdel 用户名。

(15) 建目录,并修改权限。(每个节点操作)

```
[root@yingshu1 /]# mkdir -p /oracle/app
[root@yingshu1 /]# chown -R oracle:oinstall /oracle
[root@yingshu1 /]# mkdir -p /u01/app
[root@yingshu1 /]# chown -R grid:oinstall /u01
[root@yingshu1 /]# chmod -R 775 /oracle
[root@yingshu1 /]# chmod -R 775 /u01
```

(16) 使用 vi 命令修改 grid,oracle 用户的环境变量。使用 oracle 用户 vi /home/oracle/.bash_profile 增加以下内容,注意节点二上使用 ORACLE_SID=ysdb2:

```
###########################
export ORACLE_BASE=/oracle/app
export ORACLE_HOME=/oracle/app/11.2.0/db_1
export ORACLE_SID=ysdb1
export LD_LIBRARY_PATH=$ORACLE_HOME/lib
export PATH=$ORACLE_HOME/bin:/bin:/usr/bin:/usr/sbin:/usr/local/bin:$ORACLE_HOME/OPatch
alias ss="sqlplus '/as sysdba'"
```

最后一行为 SQLPLUS 写一个别名,日后省去每次登录都要输入 sqlplus / as sysdba,直接用两个 s 代替。

使用 grid 用户 vi/home/grid/.bash_profile 增加以下内容。注意节点二上使用 ORACLE_SID=+ASM2:

```
###########################
export ORACLE_BASE=/u01/app/
export ORACLE_HOME=/u01/11.2.0/grid
export ORACLE_SID=+ASM1
exportPATH=$ORACLE_HOME/bin:/bin:/usr/bin:/usr/sbin:/usr/local/bin:$ORACLE_HOME/OPatch
```

(17) 在 Linux 下 unzip 解压安装包。解压后注意安装包的权限。

```
[root@yingshu oramed]# ll
total 4442172
-rwxrw-rw- 1 root root 1395582860 Jan 5 2014 p13390677_112040_Linux-x86-64_1of7.zip
-rw-r--r-- 1 root root 1151304589 Jan 5 2014 p13390677_112040_Linux-x86-64_2of7.zip
-rw-r--r-- 1 root root 1205251894 Jan 5 2014 p13390677_112040_Linux-x86-64_3of7.zip
```

```
[root@yingshu oramed]# unzip p13390677_112040_Linux-x86-64_1of7.zip
[root@yingshu oramed]# unzip p13390677_112040_Linux-x86-64_2of7.zip
[root@yingshu oramed]# unzip p13390677_112040_Linux-x86-64_3of7.zip
[root@yingshu oramed]# ll
total 4442180
drwxr-xr-x 7 root root       4096 Aug 27 2013 database
drwxr-xr-x 7 root root       4096 Aug 27 2013 grid
[root@yingshu oramed]# chown -R grid.oinstall grid
[root@yingshu oramed]# chown -R oracle.oinstall database
```

前两个包是数据库,必须解压至一个文件夹。第三个包是 grid 集群架构,如果是安装 RAC 必须先装 grid 集群架构。

(18) 半自动配置两节点信任关系。安装 grid Infrastructure 时,有一步是自动配置对等性。自动配置对等性当然容易,有时候会失败。我们可以手工配置对等性,我们这里演示一下半自动配置对等性的方法,很有效。

重启 sshd 服务:

```
# service sshd restart
```

从 grid 用户 su - oracle 验证 Oracle 用户密码未过期;从 Oracle 用户 su - grid 验证 grid 用户密码未过期。(两节点分别执行)

```
[root@yingshu2 dev]# su - grid
[grid@yingshu2 ~]$ su - oracle
Password:
[oracle@yingshu2 ~]$ su - grid
Password:
[grid@yingshu2 ~]$
```

使用 root 用户执行,将 yingshu1 yingshu2 替换为相应的主机名。

```
sshsetup/sshUserSetup.sh -user grid -hosts "yingshu1 yingshu2" -advanced -noPromptPassphrase
sshsetup/sshUserSetup.sh -user oracle -hosts "yingshu1 yingshu2" -advanced no-PromptPassphrase
```

根据提示要求输入回车,输入 4 次密码。如果显示 SSH verification complete,则表示配置成功。

验证方法如下。
在节点一上执行:

```
[grid@yingshu1 ~]$ ssh yingshu2 date
Thu Aug 20 15:00:18 CST 2015
```

在节点二上执行:

```
[grid@yingshu2 ~]$ ssh yingshu1 date
Thu Aug 20 15:01:10 CST 2015
```

(19) 修改资源限制。vi /etc/security/limits.conf 在文件的尾部加入以下参数：

```
oracle soft nproc 16384
oracle hard nproc 16384
oracle soft nofile 655360
oracle hard nofile 655360
oracle soft stack 10240
grid soft nproc 16384
grid hard nproc 16384
grid soft nofile 655360
grid hard nofile 655360
grid soft stack 10240
```

(20) 增加 pam_limits.so 模块。vi /etc/pam.d/login 增加以下内容：

```
# ORACLE SETTING
session required pam_limits.so
```

(21) 设置内核参数。vi /etc/sysctl.conf 增加以下设置。如果系统有默认设置，则删除默认设置。不做设置也可以，将来在安装过程中使用 runfixup 自动设置。

```
# ORACLE SETTING
fs.aio-max-nr = 4194304
fs.file-max = 6815744
kernel.shmall = 167772160
kernel.shmmax = 687194767360
kernel.shmmni = 4096
kernel.sem = 10000 40960000 10000 4096
net.ipv4.ip_local_port_range = 20000 65535
net.core.rmem_default = 262144
net.core.rmem_max = 4194304
net.core.wmem_default = 262144
net.core.wmem_max = 1048576
vm.swappiness = 10
vm.min_free_kbytes = 512MB
```

使内核参数设置生效：

```
# sysctl -p
```

(22) grid 用户使用 cluvfy 校验安装环境，此命令在解压后的安装包里。校验结果作为前期配置是否成功的参考。

```
./runcluvfy.sh stage -pre crsinst -n yingshu1,yingshu2 -verbose
```

(23) 使用 grid 用户运行 ./runInstaller 开始安装集群基础架构。

如图 4-1 所示，选择跳过软件的升级更新，我们将来手工地升级数据库。因为数据非常重要，一般不把数据库服务器接入互联网，也就无法使用在线升级。

如图 4-2 所示，选择安装和配置一个集群的基础架构。第二个选项是单节点运行基础架构，往往是为了使用它的 ASM 功能。

第4章 安装RAC数据库

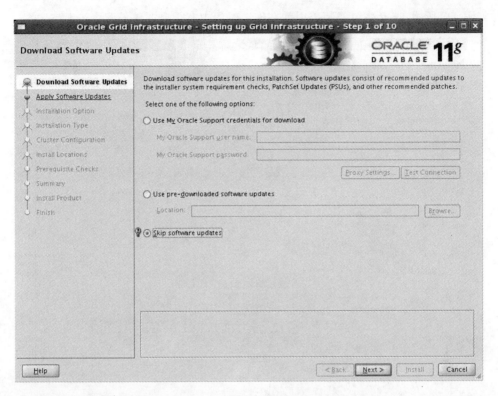

图 4-1

图 4-2

如图 4-3 所示，选择高级安装。因为在安装过程中有需要调整的配置。

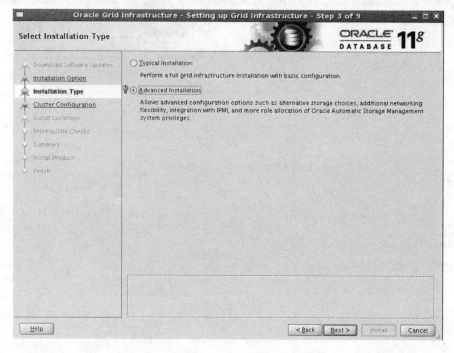

图　4-3

如图 4-4 所示，选择英语和简体中文，在将来的 Grid Control 管理中避免出现乱码。

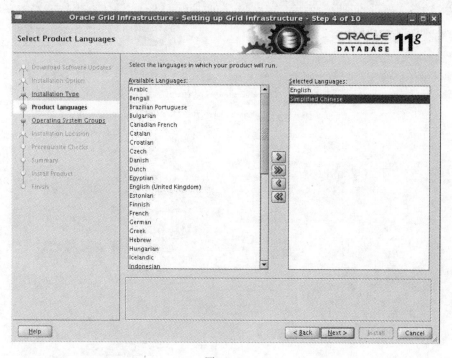

图　4-4

第4章　安装RAC数据库

如图 4-5 所示，Cluster Name 不超过 15 个字符，Scan Name 填写/etc/hosts 里配置的 scan ip 名，端口号参照整体规划。

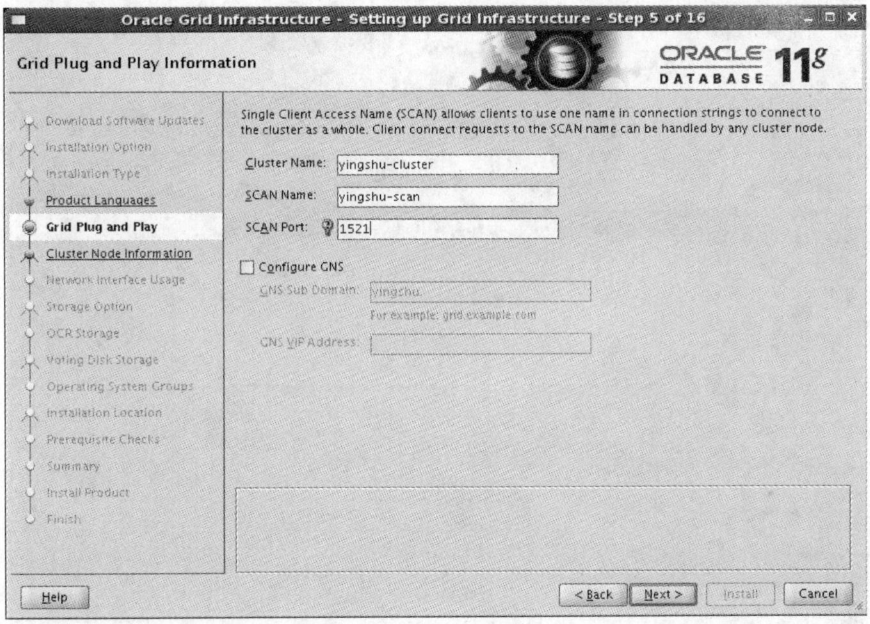

图　4-5

如图 4-6 所示，添加第二台主机名、vip 名。如果已经配置了对等性，单击 Test 按钮，测试对等性是否成功。

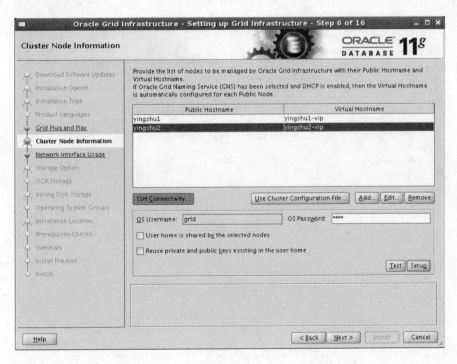

图　4-6

对等性测试成功,如图 4-7 所示。

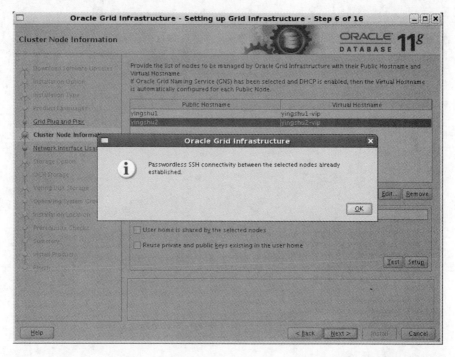

图 4-7

确认使用的网段,这里私有 IP 使用的是 10.10.10.0 网段,私有 IP 是用来做节点间数据传输用的。公共 IP 使用的是 192.168.1.0 网段,是对外公布的 IP 地址,如图 4-8 所示。

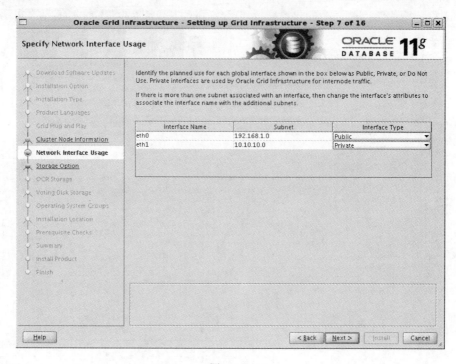

图 4-8

选择使用 ASM,如图 4-9 所示。

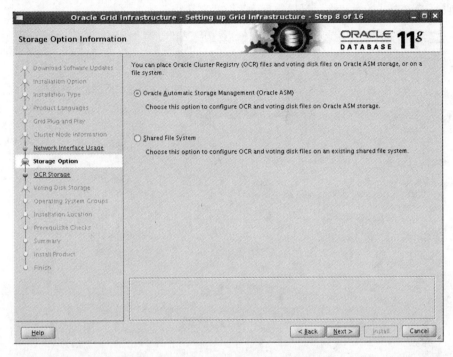

图 4-9

选择 Normal 选项,镜像 ocr、voting disk 盘。AU Size 选择 4MB,如图 4-10 所示。

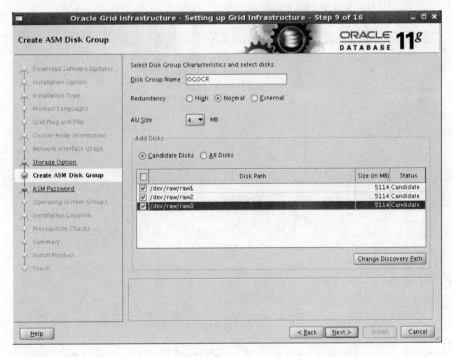

图 4-10

如图 4-11 所示，输入 ASM 密码。如果因为密码复杂程度不够收到警告，可以忽略此警告。密码复杂度要求有大小写字母、数字，且长度有要求。

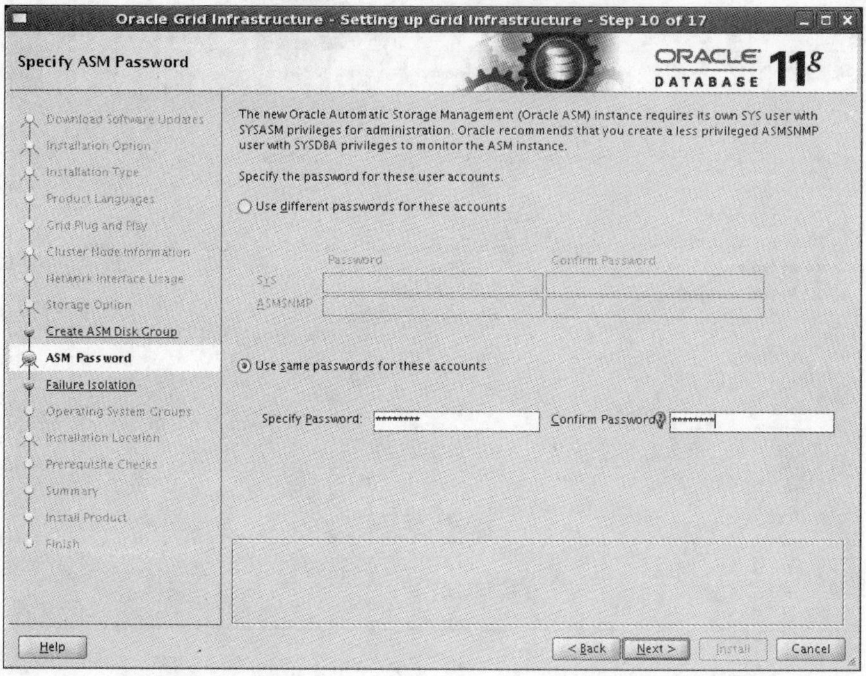

图 4-11

选择不使用 IPMI，如图 4-12 所示。

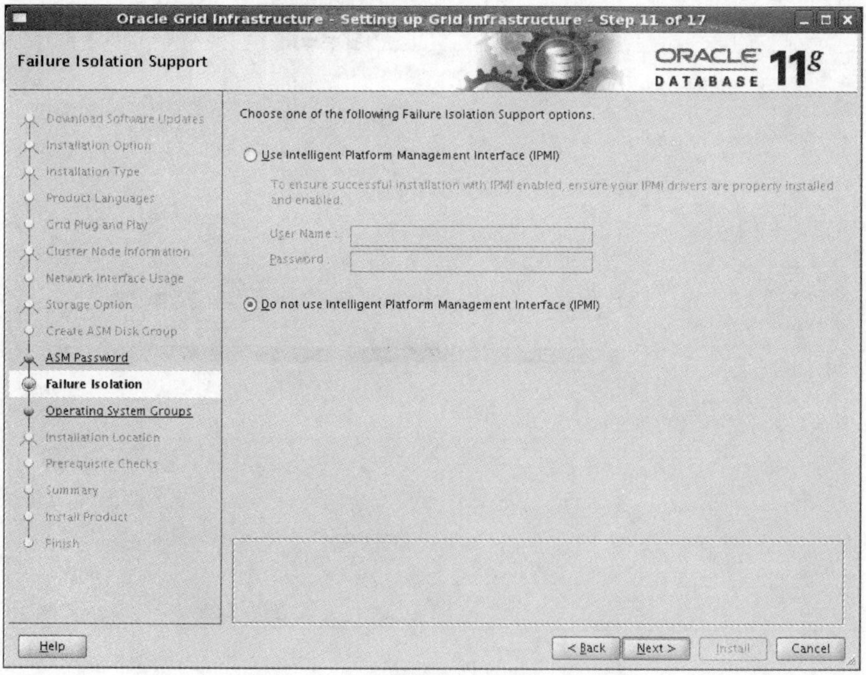

图 4-12

选择 ASM 实例、使用的用户组，如图 4-13 所示。

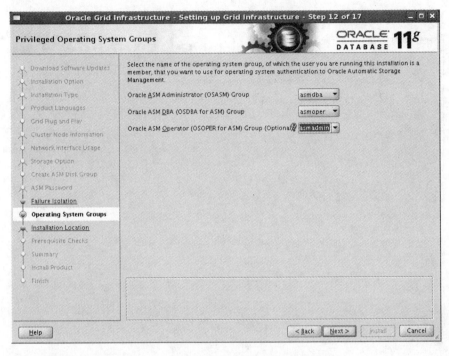

图 4-13

选择 grid 用户的 Oracle Base、Oracle Home，如图 4-14 所示。

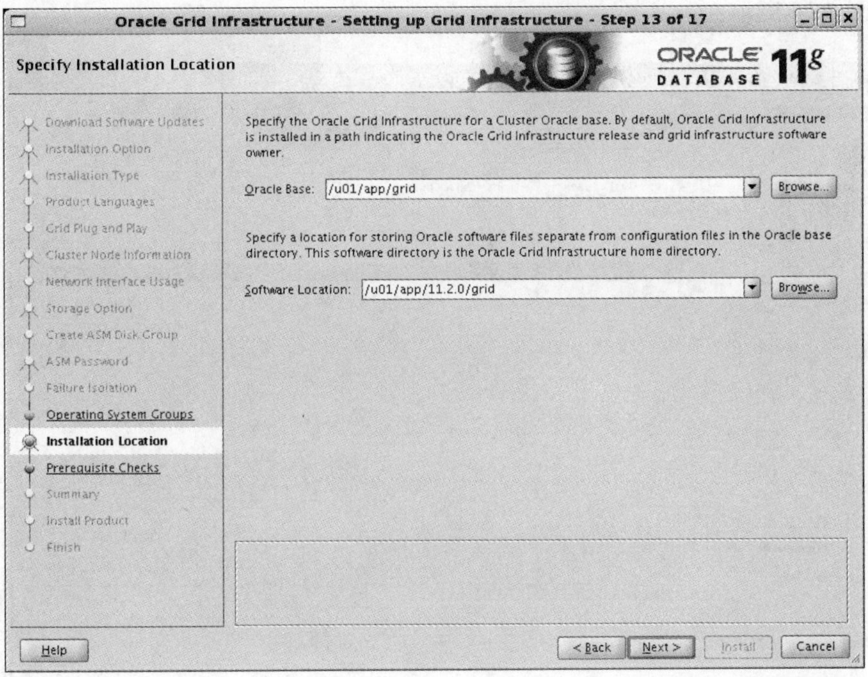

图 4-14

选择 oraInventory 路径,如图 4-15 所示。

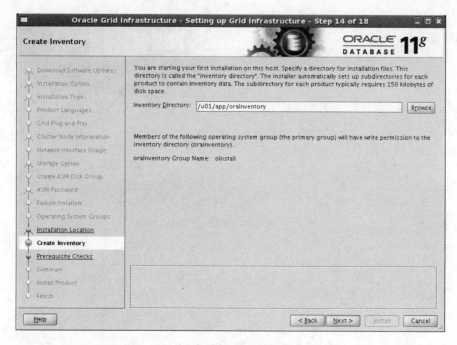

图　4-15

此页面很重要,有些 Warning 可以忽略,如操作系统参数。有些 Warning 不能忽略,如 ASM 路径。一些配置可以通过 Fix & Check Again 执行一个脚本修复好。单击 Ignore All 继续,如图 4-16 所示。

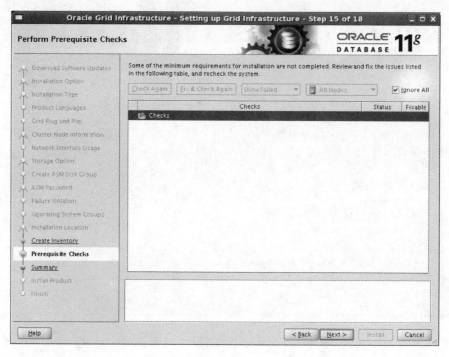

图　4-16

执行 root.sh 是里程碑，很多准备不足的问题会导致此脚本运行失败。先在第一个节点运行 orainstRoot.sh，然后运行 root.sh。再在第二个节点运行，不能同时运行，如图 4-17 和图 4-18 所示。

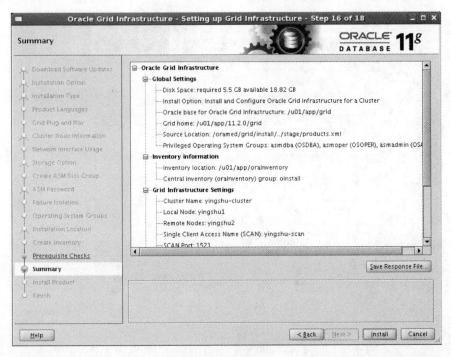

图 4-17

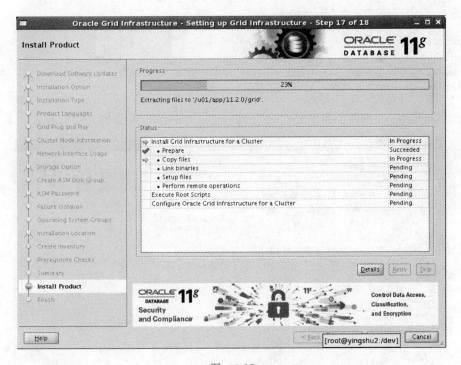

图 4-18

执行 root.sh 成功后单击 OK 按钮。进入安装检查并安装完成，INS-20802 错误可以忽略，如图 4-19 所示。

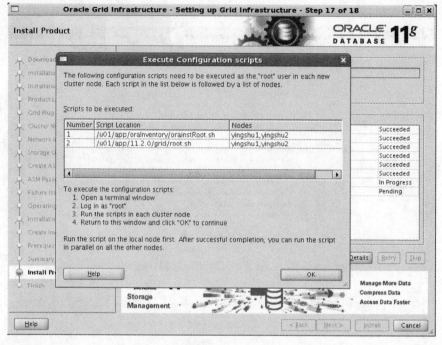

图 4-19

(24) 使用 oracle 用户运行 ./runInstaller 开始安装数据库软件。前面我们已经把软件解压好了，权限也修改为 oracle:oinstall。

```
[oracle@yingshu1 database] $ cd /oramed/database/
[oracle@yingshu1 database] $ ls
install readme.html response rpm runInstaller sshsetup stage welcome.html
[oracle@yingshu1 database] $ ./runInstaller
```

如图 4-20 所示，选择不使用 Oracle 支持的安全更新。出现警告时选择 Yes。

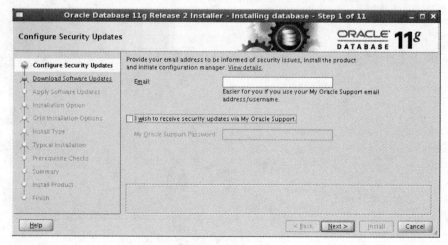

图 4-20

选择跳过软件更新，如图 4-21 所示。

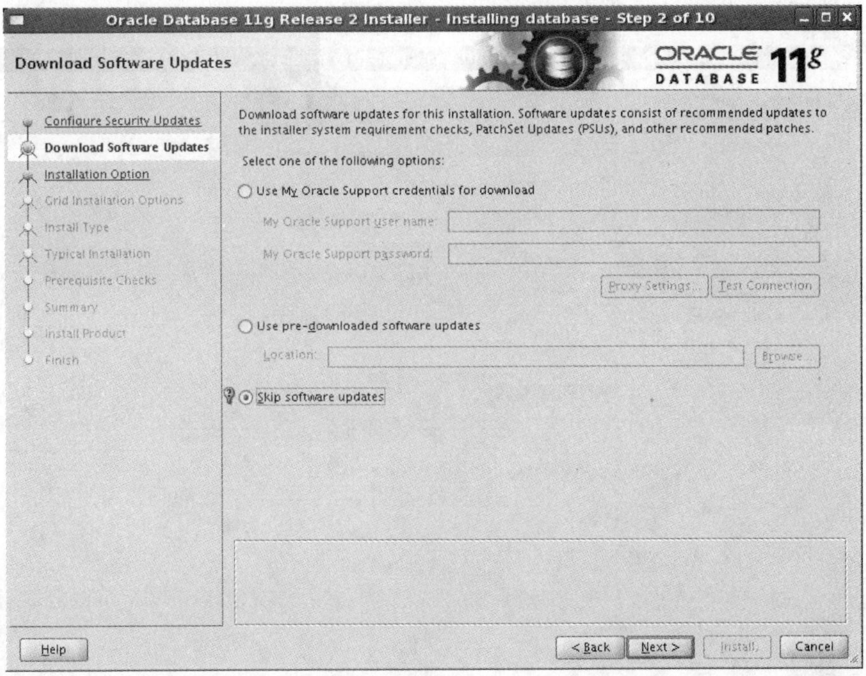

图　4-21

选择只安装软件，如图 4-22 所示。

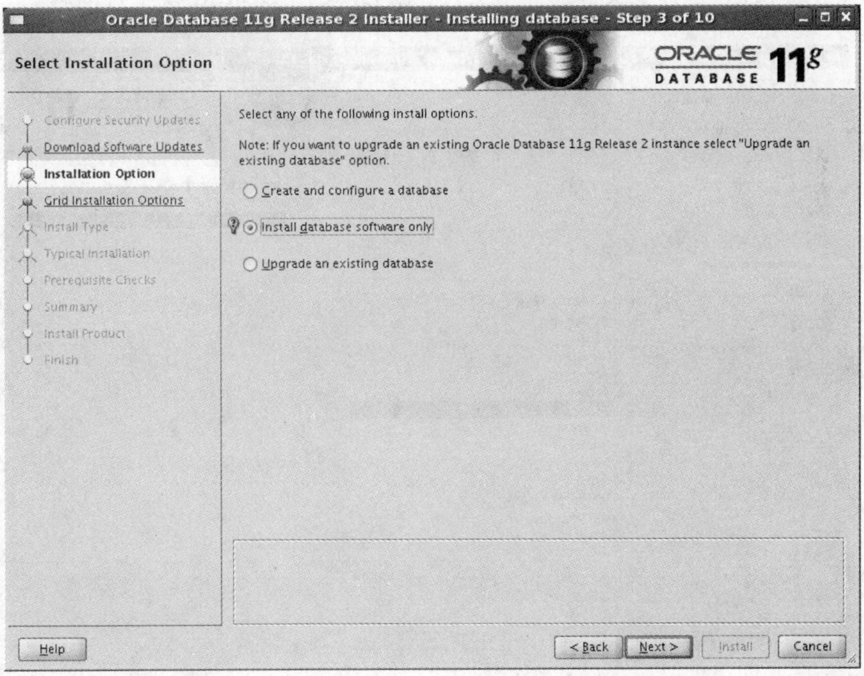

图　4-22

选择第二个选项，RAC数据库安装。测试两节点间Oracle用户的对等性，方法和gird用户对等性一样，如图4-23所示。

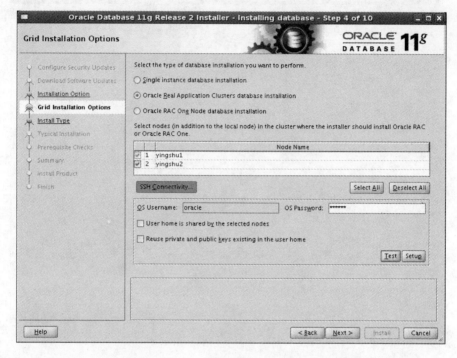

图 4-23

如图4-24所示，选择英文和简体中文，避免将来使用OEM出现字符乱码。

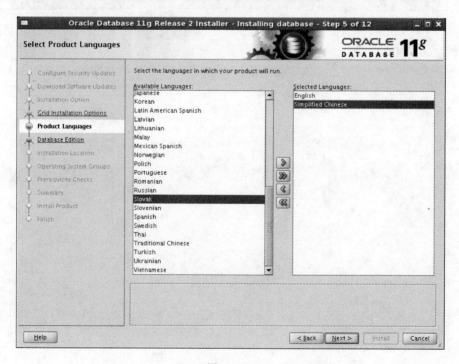

图 4-24

选择安装企业版，如图 4-25 所示。企业版中包含分区表，DataGuard 等高级功能。

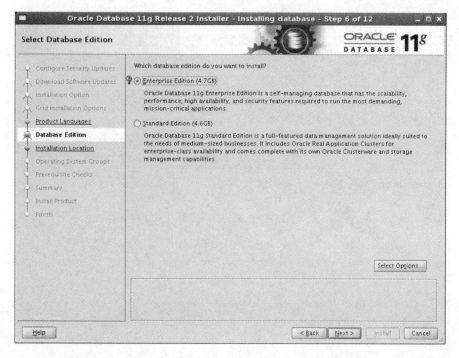

图 4-25

选择使用的功能选件，取消不使用的组件，如图 4-26 所示。

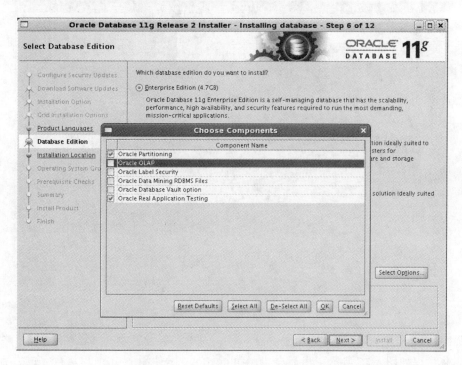

图 4-26

按照规划,填写好 Oracle Base、Oracle Home,如图 4-27 所示,环境变量设置正确,会自动弹出。

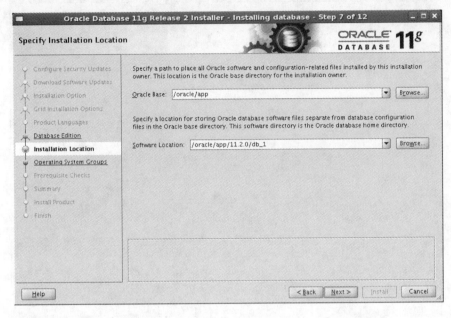

图 4-27

选择操作系统组,如图 4-28 所示。

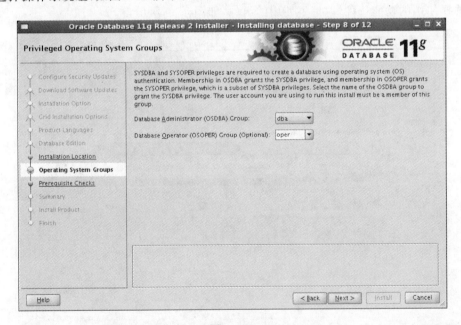

图 4-28

如图 4-29 所示,检查安装环境是否已具备。对于参数设置之类的警告,可以忽略。

安装过程如图 4-30 所示。最后会提示执行 root.sh,使用 root 用户执行完成之后,单击 Close 完成软件安装。

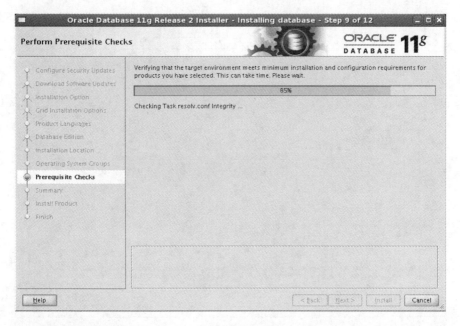

图 4-29

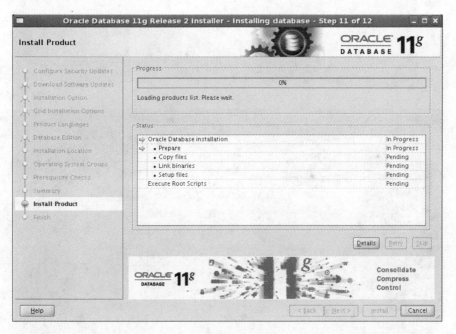

图 4-30

（25）数据库软件安装完成之后，使用 grid 用户运行 asmca，开始创建 ASM 磁盘组 dgdata。我们要在 dgdata 上建立数据库。

单击"Oreate"Disk Group，弹出图 4-31 所示的界面。选择存放数据的磁盘，选择外部冗余方式，单击 Show Advanced Options，Allocation Unit Size 选择 4MB，其他选项不动。单击 OK 按钮完成创建。

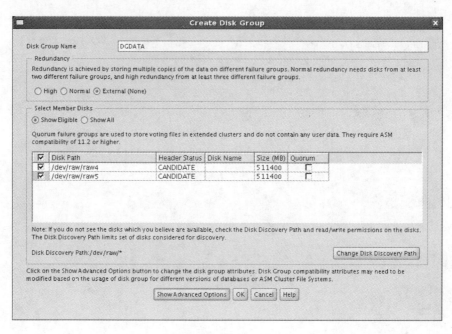

图 4-31

关于 ASM 磁盘的操作，我习惯尽量采用 asmca 来操作，和 netca 操作 listener 类似，少采用命令行。

（26）数据库软件安装完成之后，使用 oracle 用户运行 dbca 开始建库。

选择 RAC 集群数据库，如图 4-32 所示。

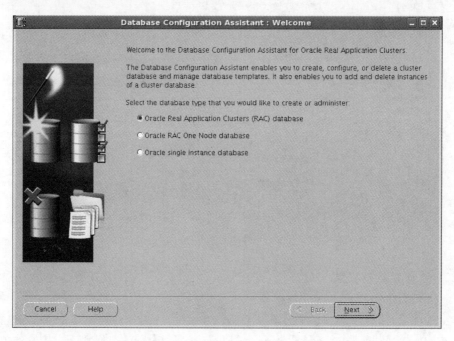

图 4-32

选择创建一个数据库,如图 4-33 所示。

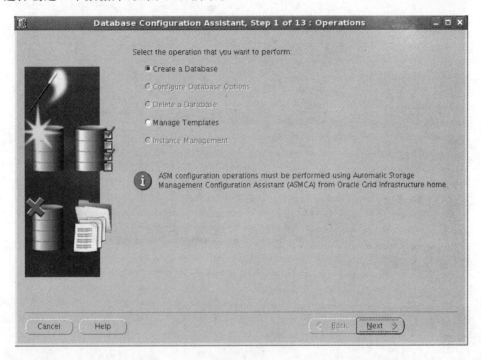

图 4-33

选择定制一个数据库,如图 4-34 所示。

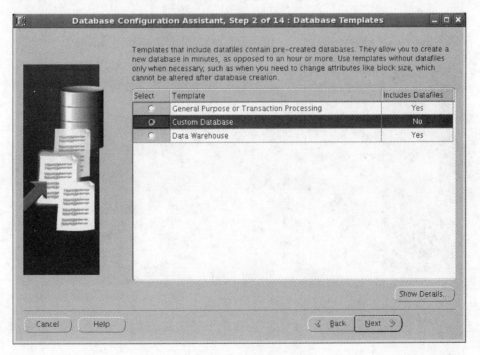

图 4-34

填写规划好的数据库名，Select All 选择所有节点，如图 4-35 所示。

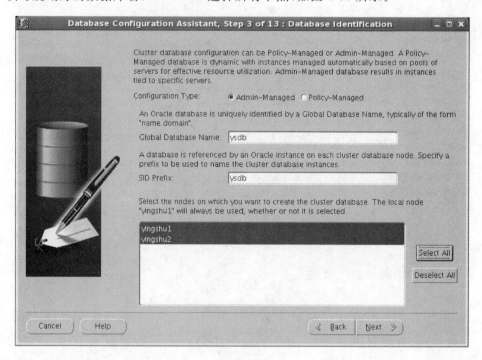

图　4-35

我们这里不使用企业管理器，如图 4-36 所示。

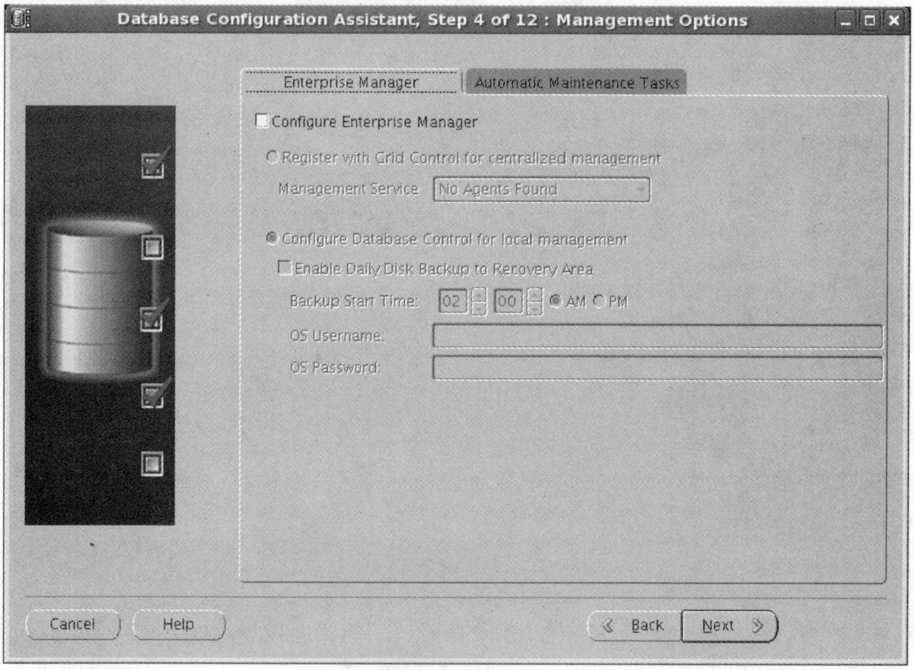

图　4-36

不使用自动维护任务。将来手工收集统计信息,手工调整性能,不使用建议报告,如图 4-37 所示。

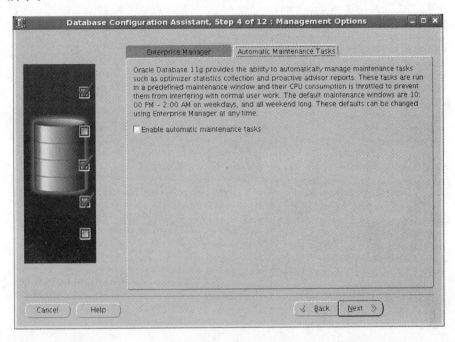

图 4-37

如图 4-38 所示,设置 sys 和 system 用户的密码。将来随时可以修改。此密码也有复杂策略要求,可以忽略。

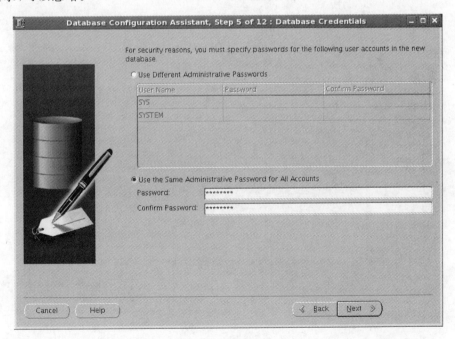

图 4-38

如图 4-39 所示，我们将数据库放在＋DGDATA 磁盘组上。这个磁盘组就是我们刚才用 asmca 创建好的磁盘组。

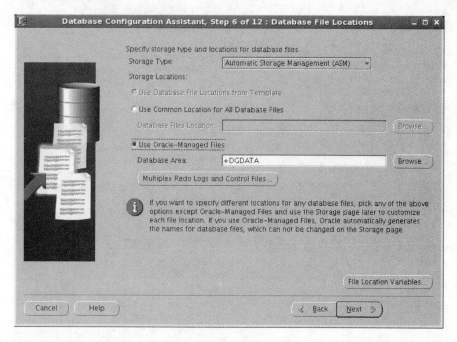

图　4-39

如图 4-40 所示，选择不设置快速恢复区，归档模式建好库之后我们手工修改。将来我们再建一个磁盘组＋DGARCH 专门存放归档日志。

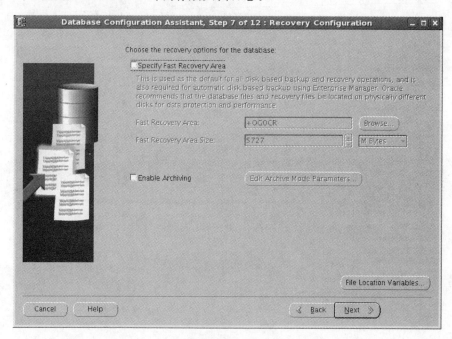

图　4-40

如果应用开发商没有要求安装某些软件,此处我们选择全部不安装,如图 4-41 所示。在以后的升级过程中也会节省更新数据字典的时间。建好库之后,也可以手动安装。

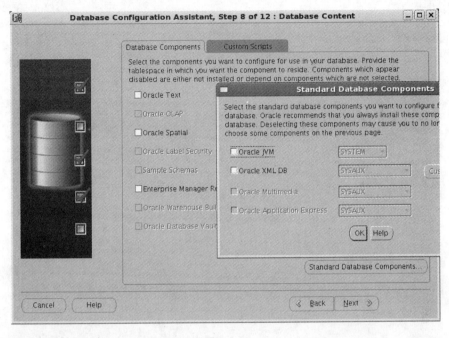

图 4-41

内存手工管理,计算好各个内存区域的大小,从稳定性角度出发,手工管理内存,如图 4-42 所示。

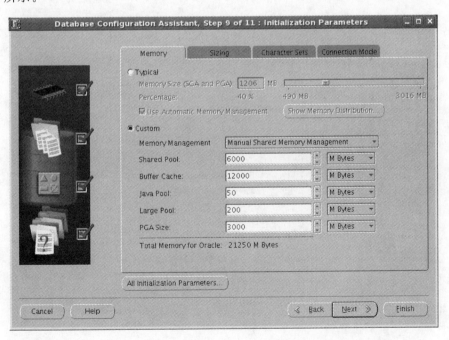

图 4-42

如图 4-43 所示，采用 8KB 的数据块，如果是 OLAP 系统可以采用 16KB 或 32KB 的数据块。设置合适的 Processes 大小，此值与数据库总的连接数有关，将来调整也可以，不过要重启数据库。

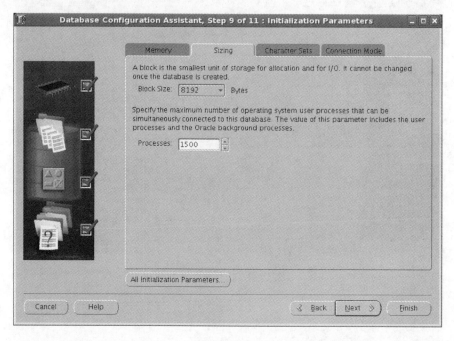

图 4-43

按照规划和应用开发商的要求，设置合适的数据库字符集，如图 4-44 所示。

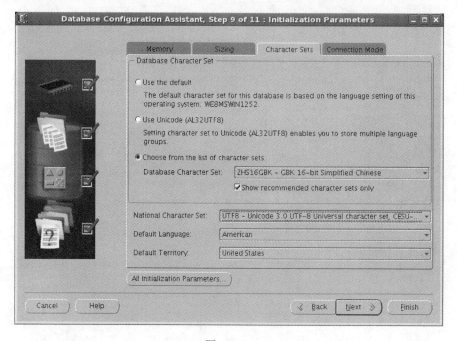

图 4-44

如图 4-45 所示，连接方式采用专用服务模式。共享服务模式会节约一些内存，牺牲一些性能，但我们对性能要求非常苛刻。

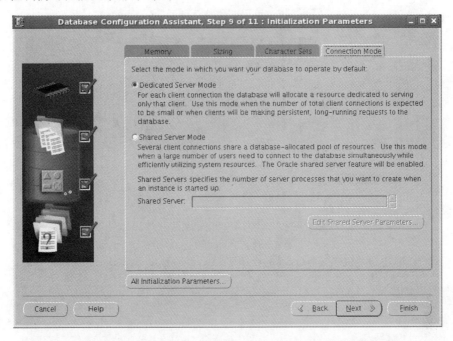

图 4-45

如图 4-46 所示，为了数据文件大小一致，管理有条理，备份恢复性能好，关闭所有数据文件自动扩展。我们将来手工添加数据文件扩展表空间。

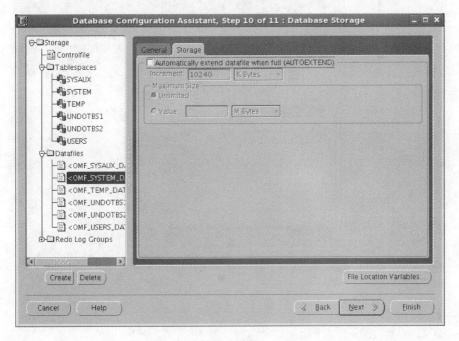

图 4-46

如图 4-47 和图 4-48 所示，统一设置数据文件的大小，尽量不要大的大，小的小，看起来也很混乱，管起来也不好管。

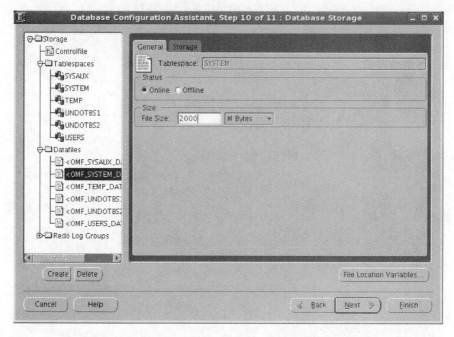

图 4-47

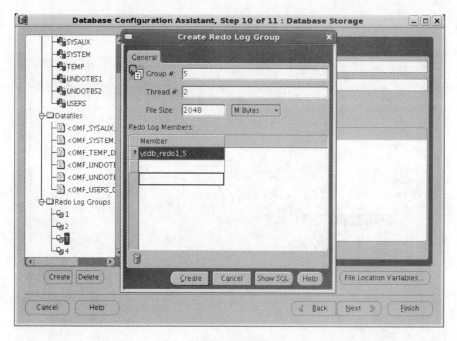

图 4-48

如图 4-49 所示，redo 日志添加至 6 组，每个实例 3 组，每组一个成员，每个成员大小为 2GB，redo 需要放在 RAID1 上，否则性能不好。

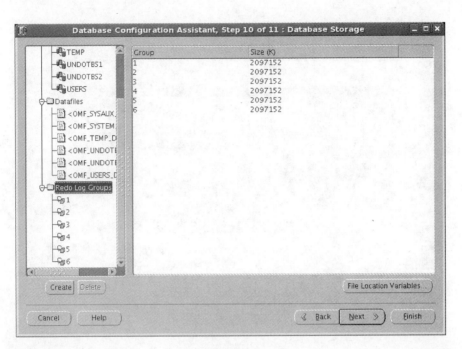

图 4-49

如图 4-50 所示，选择创建一个模板，选择生成创建脚本。单击 Finish 按钮完成，弹出摘要界面，单击 OK 按钮开始建库。

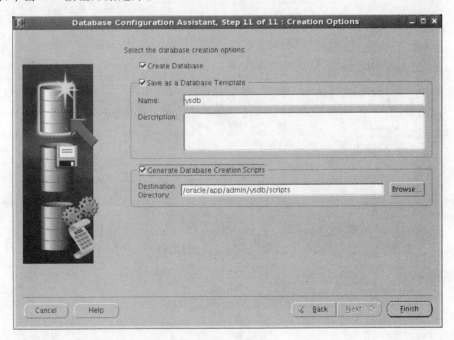

图 4-50

从图 4-51 所示的建库过程可以观察到，先建实例，然后建数据文件，再建数据字典，最后完成创建。

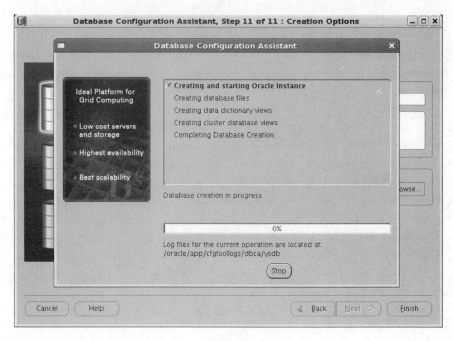

图 4-51

创建数据库完毕,如图 4-52 所示。

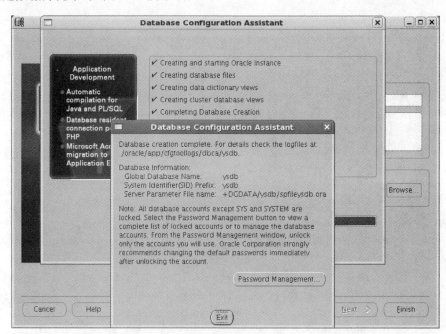

图 4-52

打补丁可以在 DBCA 建库之前,也可以在 DBCA 建库之后,打补丁的过程详见本书中的第 10 章。

第 5 章 数据库的备份和恢复

5.1 备份和恢复概述

备不备份取决于数据的重要程度,如果数据不重要,都是一些日志,又可以再造,又怕开归档影响性能,实时性要求又不高,停一天也没事儿,那可以不备份。一旦有东西坏了,重装一次软件,再建一个新库再造一份数据就行了,如测度库。

如果数据不太重要,丢一天的数据也可以接受,不想花太多的钱和精力在数据库上,数据量也不大,导出来的时间可以接受,那可以使用 exp 或 expdp 逻辑备份,每天自动执行一个导出,不开归档,省空间,省事儿。

数据库上线投产之后,基本上都很重要。有的对运维的要求很苛刻,如银行,要求 7×24 小时服务,一年也没有多长时间的停机时间,数据非常重要,一条也不能丢,所以银行不但要备份,还要容灾,两地三中心容灾,使用存储级复制、DataGuard、GoldenGate 等容灾方式。有了备份心里才踏实,定期做一次备份有效性测试,测一测能不能正常恢复,或者我们的应急系统能不能正常切换。

有的数据库晚上下班之后压力小,提前通知用户晚上可以停几个小时,数据库也很重要,比如医院。医院要重启一下数据库,提前申请晚上的停机时间,一般还是允许的。医院也要使用 RMAN 备份,然后再 DataGuard 做个容灾。

如果备份工作能在半小时左右完成,甚至几分钟十几分钟就能完成,那么可以设置成每天一个全备。也要参考每天生成的日志量。如果日志产生量很大,可能要几个小时就备一次归档日志,备完就删除,给新的归档日志留空间。

数据库备份可以根据制订的备份策略编写 RMAN 脚本,然后加到 crontab 定时任务里实现自动备份。也可以使用备份软件实现,如 Veritas Netbackup、NBU 等。NBU 不仅可以备份数据库,还可以备份应用程序和其他东西,况且集中管理,可管理多台机器,所以 NBU 卖得也不错。如果就很少的几套数据库,数据库又不大,应用变化量少,可以手工备份,那么可以只使用 RMAN 备份,不使用备份软件。

将组成数据库的操作系统文件从一处拷贝到另一处的过程叫物理备份。使用 RMAN（Recovery Manager，恢复管理器）实现，数据库必须运行在归档模式下，该方法可以实现数据库的完全恢复，做到不丢数据，需要大量的外部存储设备，例如磁带库。冷备份和热备份都叫物理备份，它不涉及逻辑内容，只对数据库中的数据块进行备份。

将数据库 shutdown immediate，然后将数据文件、控制文件、日志文件、参数文件拷贝至另一个地方的备份方式叫冷备份。冷备份只能恢复到备份的时间点，无法做完全恢复。只在允许的停机时间比较长，做数据库迁移，数据库升级的时候用得较多，其他情况下很少用这种备份方式。

数据库在运行状态下，且已经打开了归档模式，使用 RMAN 工具进行的全备或增量备份叫热备份。也就是我们最常用的备份方式，本章节将重点讨论 RMAN 热备份。热备份不需要停止数据库，但会消耗系统资源，一般在系统清闲的时间进行，如凌晨 1 点开始。

使用 SQL 语句，在数据库中抽取数据，并存为二进制文件的过程叫逻辑备份。抽取工具就是 exp 导出工具和 expdp 数据泵。逻辑备份只能恢复到开始备份的时间点，对备份之后发生改变的数据无能为力，逻辑备份常常用于做数据迁移和数据移动，GoldenGate 同步也可以使用 exp 或 expdp。

5.2　RMAN 备份的特点

5.2.1　RMAN 备份的优点

（1）支持多种恢复方法，进行完全恢复和不完全恢复。基于时间点的恢复，可以恢复到发生误操作前的时间点，将误删除的数据找回来。基于 SCN 号的恢复，精确恢复到某一个 SCN 号，在做数据同步时会用到。基于日志序列的恢复，如果归档日志文件丢失，我们只能恢复到 gap 的前一个日志，就可以使用这种恢复方法。基于退出的恢复，恢复语句的不同，恢复的终点也不一样，可以理解为截止到某一个点，就退出恢复。完全恢复，如果我们有一份全备、全备之后所有的归档日志和在线日志三样东西，我们就可以做数据库的完全恢复，做到一条数据也不丢。

（2）压缩备份，减少备份集占用的空间。我测试的 Oracle 11g 压缩备份只占不压缩备份空间的 1/3 左右。压缩备份会消耗更多的 CPU，我们在用的时候，开始的几次备份多关注一下 CPU 资源。压缩备份还会拉长备份时间，会将备份时间拉长 3 倍左右甚至更多。

（3）支持不停库备份，不需要停止数据库，不影响生产正常运行，就能实现全备份。

（4）支持并行，同时几个通道备份，对于需要加快备份速度还是很有必要的。并行备份可以设置为并行度为 4 或者 8，具体根据有多少颗 CPU，一般不超过 CPU 数的一半，还要考虑存储的速度，如果存储跟不上，再并行也不快。

（5）可以设置成自动任务，不需要人工干预，我们只要经常关注一下备份日志，查看备份是否成功了就可以了。自动任务让我们想什么时候备份都可以。

（6）支持多种备份类型，RMAN 热备可以备份数据文件、控制文件、归档文件、参数文件，但不能备份在线的 redo 文件。要保护好在线 redo，可以给 redo 所在的磁盘做一个 RAID1 镜像，也可以设置每组两个 member，分别放到不同的磁盘上，坏一个 member 没有关系。备份类型还包括全库备份和增量备份,如果数据库很大,几十 TB,可以使用增量备份,缩短备份时间。每周日全备,周一到周六增量备份。

(7)支持跨平台的表空间迁移,可以使用 rman convert 命令实现跨平台的数据迁移。

(8) RMAN 功能强大,具有很强的脚本控制语言,可以根据我们的需求,编出合适的备份恢复脚本。

5.2.2 RMAN 备份的缺点

(1)脚本编写比较复杂。功能越强大对应的脚本越多,像买一个高级的音响或电视机,功能强大的一辈子也用不上几回,RMAN 也有这么强大的功能。作为专业人士,我们要知道有这些功能,一旦有这样的工作,我们得会用。

(2)占用较大的存储空间。一份数据,为了备份容灾要存好几份它的副本,如果生产存储上做的是 RAID10,这就 2 份;再备份,备份存储上也是做的 RAID10,这就 4 份,再保留两份的备份集,这就 6 份。再做一个容灾,不管是同城的还是异地的,又会多至少一份,容灾的存储如果再是 RAID10,那就 8 份了;再两地三中心……总共算下来,一共七八倍的冗余。安全倒是安全了,火灾地震都不怕,但确实浪费了不少存储资源。我们要在冗余、安全和性能上做一个很好的平衡。既够用,且安全,又节约,以此来指导我们的规划工作。

(3)要开启归档模式。归档模式开了之后,务必想着定期删除归档日志,否则迟早会满。有的客户 DBA 责任心不强,也不买服务,存储够大,堆了两年的归档日志,终于满了,数据库卡住了,还着急找原因。

5.3 RMAN 的配置

5.3.1 非归档切换到归档日志模式

设置归档日志的路径,我们将归档日志放在什么地方:

```
SQL> alter system set LOG_ARCHIVE_DEST_1 = 'LOCATION = /arch1' scope = both;

System altered.
```

使用 LOG_ARCHIVE_DEST 参数也可以,但只能归到本地;使用 LOG_ARCHIVE_DEST_1 既可以归到本地也可以归到远程。其他参数如归档日志的格式 log_archive_format,归档日志进程的最大值 log_archive_max_processes 使用默认值就可以。在 Oracle 10g 之前我们还要修改 log_archive_start 参数,Oracle 10g 和 Oracle 10g 之后,这个参数就被淘汰了。

非归档改为归档模式,必须重启数据库,在 mount 状态下修改归档模式。步骤如下:

```
SQL> shutdown immediate;                          //关闭数据库
Database closed.
Database dismounted.
ORACLE instance shut down.
SQL>
SQL> startup mount;                               //将数据库启动到 mount 状态
```

```
ORACLE instance started.
... ...
Database mounted.
SQL>
SQL> alter database archivelog;              //开启归档模式
Database altered.
SQL> alter database open;                    //打开数据库
Database altered.
SQL>
SQL> archive log list;                       //查看是否归档,归到什么地方
Database log mode              Archive Mode
Automatic archival             Enabled
Archive destination            /arch1
Oldest online log sequence     25
Next log sequence to archive   27
Current log sequence           27
```

5.3.2 RMAN 的配置

登录到 RMAN 之后,使用 show all 命令,可以查看当前数据库的 RMAN 配置。在这里,也可以修改 RMAN 配置。如配置保留备份集的策略,是否压缩,控制文件自动备份,使用磁盘/磁带设备类型,是否开启备份优化功能等。以下是常用的配置方法。

1. show all 查询当前的数据库 RMAN 配置

```
[oracle@yingshu ~]$ rman target /
connected to target database: YSDB (DBID = 3736627791)
RMAN> show all;
using target database control file instead of recovery catalog
RMAN configuration parameters for database with db_unique_name YSDB are:
CONFIGURE RETENTION POLICY TO REDUNDANCY 1; # default
CONFIGURE BACKUP OPTIMIZATION OFF; # default
CONFIGURE DEFAULT DEVICE TYPE TO DISK; # default
CONFIGURE CONTROLFILE AUTOBACKUP OFF; # default
CONFIGURE CONTROLFILE AUTOBACKUP FORMAT FOR DEVICE TYPE DISK TO '%F'; # default
CONFIGURE DEVICE TYPE DISK PARALLELISM 1 BACKUP TYPE TO BACKUPSET; # default
... ...
CONFIGURE COMPRESSION ALGORITHM 'BASIC' AS OF RELEASE 'DEFAULT' OPTIMIZE FOR LOAD TRUE ; # default
```

默认的配置可能不适合我的备份策略,我们在给数据库做备份之前,需要修改一下 RMAN 的配置。我们在制订备份策略的时候,也可能包含修改 RMAN 的配置。

2. 设置控制文件自动备份

```
RMAN> configure controlfileautobackup on;
new RMAN configuration parameters are successfully stored

RMAN> configure controlfileautobackup format for device type disk to '/orabak/ysdb_ctl_%f';
new RMAN configuration parameters are successfully stored
```

当发生数据库备份时,或建表空间,删除 log 文件等物理结构发生改变时,Oracle 会自

动备份控制文件。Oracle 10g 会立刻自动备份，Oracle 11g 会有几分钟的延迟，具体延迟时间受参数"_controlfile_autobackup_delay"的控制。默认是 300 秒。

查询这个隐含参数的方法：

```
SQL> col ksppinm format a54
SQL> col ksppstvl format a54
SQL> select ksppinm, ksppstvl
  2  from x$ksppi pi, x$ksppcv cv
  3  where cv.indx = pi.indx and pi.ksppinm like '\_%' escape '\' and pi.ksppinm like '%_controlfile_autobackup_delay%';

KSPPINM                                          KSPPSTVL
------------------------------------------------ --------
_controlfile_autobackup_delay                    300
```

3. 设置备份集的冗余

备份经常进行，一天一次甚至每隔几小时备一次，有了最新的备份，前几天的备份就可以删除，省得它占用宝贵的存储资源。删除之前我们可以设置一个策略，先将它标记为废弃 obsolete，例如我们保留最近的三份全备，或设置成保留 7 天之内的备份，7 天以前的备份标记为 obsolete。默认的冗余是 1，保留一份备份。

将冗余策略改为保留 2 份：

```
RMAN> configure retention policy to redundancy 2;
new RMAN configuration parameters are successfully stored
```

将冗余策略改为保留 7 天。保留策略只能配置一种，新的配置保留 7 天会将上面配置的保留 2 份覆盖掉。两种策略只能有一种生效。

```
RMAN> configure retention policy to recovery window of 7 days;
old RMAN configuration parameters:
CONFIGURE RETENTION POLICY TO REDUNDANCY 2;
new RMAN configuration parameters:
CONFIGURE RETENTION POLICY TO RECOVERY WINDOW OF 7 DAYS;
new RMAN configuration parameters are successfully stored
```

4. 删除过期备份

obsolete(废弃)标记在执行 REPORT OBSOLETE 或者是 DELETE OBSOLETE 时用到。标记为 obsolete 的备份并没有被物理删除，它在物理上是存在的。是可以再用作恢复的。它取决于我们的保留策略，如果我们的保留策略是保留 7 天，那么 7 天前的备份将标记为 obsolete。obsolete 的意思是不需要了，有更新的了。和 obsolete 相关的命令是：

```
report obsolete;                    //报告废弃的备份
delete noprompt obsolete;           //删除废弃的备份,物理删除
```

expired(过期)，标记为过期的备份在物理上已经不存在了，只是在控制文件中还记录着这个备份文件的信息。可以使用 crosscheck backup; 命令查看是否有已经被物理删除的备份集。用 delete expired backup; 删除过期的记录。expired 的意思是找不到了，不存在

了。和 expired 相关的命令是：

```
crosscheck archivelog all;                    //校验是否有过期的归档日志备份
delete noprompt expired archivelog all;       //删除过期归档日志备份,不提示
crosscheck backup;                            //校验是否有过期备份集
delete noprompt expired backup;               //删除过期备份,不提示
```

实现自动删除，将上述的删除语句添加到 RMAN 的自动备份脚本里，备份和删除过期备份先后执行。或将物理删除的脚本直接加到 crontab 里，也可以实现自动删除。

5. 配置并行度

在数据库备份时，并行分配几个通道。默认是 1 个。如果 CPU 是 4 颗 16 核，我们为了提高备份速度，可以将并行度设置成 4。自动分配 4 个通道进行备份。

```
RMAN> configure device type disk parallelism 4 backup type to backupset;
new RMAN configuration parameters:
CONFIGURE DEVICE TYPE DISK PARALLELISM 4 BACKUP TYPE TO BACKUPSET;
new RMAN configuration parameters are successfully stored
```

6. 配置备份优化功能

默认是关闭。开启后，在特定的条件下，备份将跳过已经在备份集中存在的文件，以减少备份时间，节约空间。思路就是能不备份的文件就不备份了。什么情况下数据文件才可以不备份呢？同样的修改时间，同样的 SCN 号时就可以不备份。

当数据文件被 offline-normal、数据文件 read-only 或者是 closed normally 后，备份过了，新备份发起时，这些数据文件都没有发生任何变化，它和备份集里的数据文件拥用同样的 SCN 号，且最后的修改时间相同，就可以不用备份了。

归档日志文件，当备份集里面已经有这个归档日志了，那就可以不重复备份了。官方文档上说的是要有同样的 DBID、thread、sequence number、RESETLOGS SCN 和 time 修改时间。

```
RMAN> configure backup optimization on;                    //打开备份优化功能
new RMAN configuration parameters are successfully stored
RMAN> configure backup optimization off;                   //关闭备份优化功能
new RMAN configuration parameters are successfully stored
```

7. 配置备份片的最大值

如果数据库很大，有几十个 TB，我们可以把一个备份片也让它长到几个 TB，要给备份片设一个最大值，达到这个最大值的时候，另外生成一个新的文件。

```
RMAN> configure channel device type disk maxpiecesize 10g;
new RMAN configuration parameters are successfully stored
……
released channel: ORA_DISK_4
```

8. 配置 RMAN 的备份压缩比例

如果有兴趣，可以做一下不同压缩比例设置下做全备份，备份集的大小和备份的时间分别有多大区别。常用的压缩模式是 medium，或者是默认值 basic。

```
configure compression algorithm 'HIGH';        //高度压缩,速度最慢
configure compression algorithm 'MEDIUM';      //中度压缩,速度比 low 慢
configure compression algorithm 'LOW';         //最小的压缩比,速度最快
configure compression algorithm 'BASIC';       //压缩比和 medium 差不多,但速度慢
```

9. 配置归档日志删除策略

```
configurearchivelog deletion policy to applied on all standby;
```

在有 DataGuard 的情况下,如果备库还没有应用归档日志,主库就不可以使用 DELETE INPUT 删除归档日志。

```
configurearchivelog deletion policy to backed up 1 times to disk;
```

确认归档日志已经在磁盘上备份了一份了,那么就可以在备份脚本里添加 DELETE INPUT 语句进行删除了,否则不允许删除。

10. 配置控制文件快照

```
configure snapshot controlfile name to '/oracle/app/oracle/product/10.2.0/db_1/dbs/snapcf_ysdb.f'; # default
```

默认的快照是放在 ORACLE_HOME/dbs 底下,我们一般不用修改它的位置。控制文件快照在备份控制文件时会用到,可以用 RMAN 将其还原(restore)成控制文件。如果控制文件丢失了,请记得这里还有一个快照呢。

5.4 数据库全备

针对数据库比较小,备份时间短的数据库,可以使用数据库全备的方法,实现每天全备。方法最简单,效率高,脚本虽然简单,但很实用,当然了,前提是归档模式。对应脚本如下:

```
RMAN > run {
backup full tag 'ysdb' database
include current controlfile format '/orabak/ysdbfull_%d_%T_%s'
plusarchivelog format '/orabak/arch_%d_%T_%s' delete all input;
}
```

过程是第一步先分配通道,通道的数量是我们在 RMAN 中配置的数量,也可以手工分配通道,通道的数量影响备份速度;第二步是备份归档日志文件,备完之后删除归档;第三步是备份数据文件,相当于拷贝;第四步是切换日志,将在线日志归档,备份过程中产生的新归档日志这时开始备份,然后删除;第五步是备份控制文件,备份 spfile。

5.5 备份部分内容

备份表空间,只备份一个表空间:

```
RMAN > backup tablespace users format '/orabak/users_%t%s.%p';
```

只备份一个数据文件：

```
RMAN> backup datafile '+DGDATA/ysdb/datafile/users.259.885903335' format '/orabak/users%t%s.%p';
```

只备份当前的控制文件：

```
RMAN> backup current controlfile format '/orabak/control%t%s.%p';
```

只备份归档日志文件，或备份之后删除归档日志，减少空间占用：

```
RMAN> backup archivelog all format '/orabak/arc_%t%s.%p';
RMAN> backup archivelog all format '/orabak/arc_%t%s.%p' delete input;
```

5.6 增量备份

自 Oracle 10g 以来，增量备份具有真正的意义了，只备份变化的数据块，减少每次备份的数据量，可以缩短备份时间。Oracle 9i 因为要扫描所有的数据块，才能知道哪些数据块发生了变化，时间上、性能上都没有什么优势，使用较少。

增量备份分两种方式，Differential 差异增量和 Cumulative 累积增量，默认为 Differential 差异增量。它们备份的内容有所区别，如果只使用 0 级和 1 级的备份方法，区别如下。

Differential 差异增量：以上一次的备份结束为起点至目前的数据变化。

Cumulative 累积增量：每一次都是以全备（0 级备份）开始，至目前的变化。

0 级备份脚本。相当于数据库全备，能作为增量备份的基础：

```
run {
backup incremental level 0 tag 'ysdb' format '/orabak/ysdb0%u_%s_%p' database;
backup format '/orabak/arch%u_%s_%p' archivelog all delete input;
}
```

Oracle 11g 共有 5 个级别的增量备份，分别是 0、1、2、3、4。最常用的是 0 级和 1 级。下面是差异增量的脚本：

```
backup incremental level 1 database;
backup incremental level 2 database;
backup incremental level 3 database;
backup incremental level 4 database;
```

累积增量的样例：

```
backup incremental level 1 cumulative database;
……
```

5.7 查看备份情况

备份完成之后，我们可以查看备份的情况。list 命令是在控制文件或恢复目录中查询历史的备份情况。list backup summary 是列出备份的概述。

```
RMAN> list backup summary;
List of Backups
== == == == == == == =
Key     TY LV S Device Type Completion Time # Pieces # Copies Compressed Tag
----    ------------------ --------------- -------- -------- ---------- ---
1       B  F  X DISK        23-JUN-15       1        1        NO         TAG20150623T151524
2       B  A  A DISK        08-JUL-15       1        1        NO         YSDB
... ...
```

列出数据库前化身,如果数据库多次被使用 resetlogs 打开,这里会有多条记录。数据库被 resetlogs 之后,代表一个 RMAN 备份时代的结束。

```
RMAN> list incarnation ;
List of Database Incarnations
DB Key  Inc Key  DB Name  DB ID        STATUS    Reset SCN   Reset Time
-----   -----    ------   ----------   -------   ---------   ----------
1       1        YSDB     3736627791   PARENT    1           08-JUN-15
2       2        YSDB     3736627791   CURRENT   379611      21-JUL-15
```

如果还想使用 resetlogs 之前的备份,跨越 resetlogs。就需要使用 incarnation 来穿越。这样我们就可以使用(incarnation)化身号为 1 的备份,把数据库恢复至 resetlogs 之前更早的时间点。

```
reset database to incarnation 1;
```

显示备份的某个数据文件及其他相关的 list 命令。

```
list backup of datafile '+DGDATA/ysdb/datafile/system.260.881852381';
RMAN> list backup;                              列出详细备份
RMAN> list expired backup;                      列出过期备份
RMAN> list backup of database;                  列出所有数据文件的备份集
RMAN> list backup of tablespace user01;         列出特定表空间的所有数据文件备份集
RMAN> list backup of controlfile;               列出控制文件备份集
RMAN> list backup of archivelog all;            列出归档日志备份集详细信息
RMAN> list archivelog all;                      列出归档日志备份集简要信息
RMAN> list backup of spfile;                    列出 SPFILE 备份集
RMAN> list copy of datafile 5;                  列出数据文件映像副本
RMAN> list copy of controlfile;                 列出控制文件映像副本
RMAN> list copy of archivelog all;              列出归档日志映像副本
RMAN> list incarnation of database;             列出对应物/列出数据库副本
```

5.8 数据库恢复

数据库恢复主要应用于以下几种情形:❶数据库无法启动,强制启动也无济于事,忽然存储掉电可能会出现这种情况;❷生产库所在的磁盘坏了;❸配置 DataGuard 的过程中;❹挽回误操作;❺数据库迁移过程;❻数据库丢失文件。数据库最坏的情况就是坏块和无法启动之类的事。还有的数据库实在启动不了了,又发现没有备份,最后逼到用 dul 做恢复,很惨。有了备份,又做过恢复的测试,心里就踏实了。万一出现故障,我们可以轻松地再造一个新库出来。

恢复的类型:❶完全恢复;❷基于时间点的恢复;❸基于 Cancel 的恢复;❹基于 SCN 号的恢复。后三种都是不完全恢复。第一种可以做到一条数据也不丢,恢复到最后一个提交。

1. restore database

restore database 相当于物理拷贝,将备份集中的数据文件 cp 回来。restore database 是在数据库 mount 的状态下执行的,根据控制文件记录的备份信息,将最后一次成功备份集里的数据文件恢复至对应的位置。只需要这一个简单的语句,整个数据库中的数据文件就神奇地重现在原来的位置,丢失了的数据文件又回来了。过程不需要人工干预,我们只要查看一下日志是否恢复成功就行了。

```
RMAN> restore database;
Starting restore at 09-JUL-15
… …
channel ORA_DISK_1: starting datafile backup set restore
channel ORA_DISK_1: specifying datafile(s) to restore from backup set
channel ORA_DISK_1: restoring datafile 00001 to +DGDATA/ysdb/datafile/system.260.881852381
… …
channel ORA_DISK_2: restore complete, elapsed time: 00:00:38
Finished restore at 09-JUL-15
```

还可以在恢复日志中显示摘要:

```
restore database preview summary;
```

restore database 完成之后,数据库现在是一个不一致的状态,还原的数据文件是备份时的状态,早于当前的时间,数据库还启动不起来,还需要对数据库做 recover 应用日志,做完全恢复或恢复到某一个特定的状态下。

2. 各种 recover database

recover database 就是将归档日志或在线日志应用于当前的数据文件,做数据库的恢复,将数据库从时间 A 恢复至时间 B。可以是完全恢复,也可以是不完全恢复,具体要根据实际情况来选择恢复的方法。

如果归档日志和在线日志都完好无损,restore 完成之后,就可以做 recover 完全恢复数据库,一条数据也不丢,提交的数据没有任何损失。

```
RMAN> recover database;
```

基于时间的恢复:

```
RMAN> run
{
set until time "to_date('2015-07-21 15:40:00','yyyy-mm-dd hh24:mi:ss')";
recover database;
}
```

基于退出的恢复：

```
SQL> conn /as sysdba
Connected.
SQL> recover database until cancel;
ORA-00279: change 359163 generated at 07/21/2015 13:09:58 needed for thread 1
ORA-00289: suggestion : /arch1/1_39_881852367.dbf
ORA-00280: change 359163 for thread 1 is in sequence #39
Specify log: {<RET>=suggested | filename | AUTO | CANCEL}
auto
```

基于 SCN 号的恢复：

```
RMAN> recover database until scn 12345111;
```

基于日志序列号的恢复，这里请注意，日志序号为 24 的归档日志是不被恢复的。

```
RMAN> recover database until sequence 24;
Starting recover at 24-JUL-15
using channel ORA_DISK_1
starting media recovery
```

或者 RAC 环境使用 thread 号：

```
RMAN> recover database until sequence 24 thread 1;
```

以上的不完全恢复，不能以 alter database open 直接打开数据库，必须使用 resetlogs 或者 noresetlogs 选项打开数据库。在线日志完好的情况下，noresetlogs 将 recover 恢复在线日志，不丢数据。而 resetlogs 将不恢复在线日志内容，导致在线日志中的数据丢失，重新启用新的日志序列号，以前的数据库备份宣告截止。建议重新做一次全备。

```
SQL> alter database open noresetlogs;
Database altered.
SQL> alter database open resetlogs;
Database altered.
```

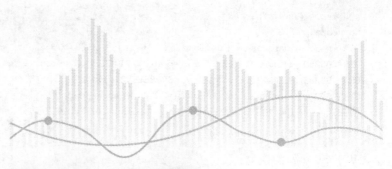

第 6 章
数据库故障处理

6.1　故障处理概述

　　故障处理的工作是不是好活儿,要看运气。如果碰见很浅显的问题,那么三下五除二就药到病除了,几分钟的事儿。如果碰见几样难办的问题,那谁碰上谁倒霉,刀架脖子上好几天也不敢说完全恢复。一般来说,数据库坏块,数据库死活启动不起来,数据丢失又没有备份算最惨的情况,碰上这三种情况,那考验我们的时候就到了,很可能要影响生产,造成损失,此类为高级别严重的事件。故障的原因五花八门,情况各异,本书中不可能将其完全涉及,只能总结一些常见的故障,按故障的性质和根源进行分门别类,基本上我们碰见的问题,都在这些类别里。服务工作做久了,碰见的问题虽然很多,但大都是老一套,同样的问题在这个客户存在,那个客户也同样有。我们的目的不是处理故障显身手,而是避免故障的发生,让故障少发生甚至不发生,即使发生了故障也能有应急方案快速正确地解决,不影响生产,不给公司带来损失,这才是一个好的管理员。

6.2　最常见的一些故障

　　(1) 文件系统空间满了。可能被归档日志、trace 文件、log 文件、数据文件等撑满了。
　　(2) 表空间满了。长期增长,自动扩展到了 32GB 极限,数据库无人看守维护。
　　(3) 碰见 bug 了。使用的版本太低,碰见 bug 在所难免,最好升级。
　　(4) 达到参数设置的限额了。数据库参数、操作系统参数有最大值,超过就出问题。
　　(5) SQL 运行不出来。可能是统计信息错了或索引问题,也有的需要 SQL 调优。
　　(6) 统计信息错了。统计信息影响执行计划,要始终使用正确的统计信息。
　　(7) 误操作。误删除了表,误删除了文件,工具的原因,开的窗口太多了、混了。
　　(8) 硬件坏了,数据丢了。存储掉电重启后库起不来,主机坏、磁盘坏、网络坏等。
　　(9) 安装时该调整的某些项目没有调整到位,为日后运行埋下的隐患。

(10)数据库锁、闩、死锁。
(11)CPU 利用率过高,看看是谁占的。
(12)RAC 交叉访问太多,需要分开应用,避免交叉访问。
(13)参数设置错误,内存不够用,使用了错误的 spfile。

6.3 案例 1 存储损坏

表 6-1 为本节要讲述的案例 1。

表 6-1 案例 1

故障描述	内容	解决结果	潜在影响及后续计划
存储损坏了 5 块磁盘,导致 Oracle 数据库已丢失。需要对数据库进行恢复。数据库已经使用 NBU 进行备份,最后的备份集是 9 月 18 日,归档日志也只到 9 月 18 日,所以只能恢复到 9 月 18 日的数据	控制文件丢失,redo 日志丢失,数据文件丢失。使用 RMAN 进行数据恢复	数据库已恢复至 9 月 18 日	注意巡检时,验证数据库备份是否已经成功。更换更可靠的存储设备

数据恢复过程参照以下过程:
先把数据库启动到 nomount 状态。恢复控制文件:

```
$ rman target /
RMAN > restore controlfile to '/u01/oradata/yingshudb/control01.ctl' from '/orabak/yingshudbfull_YINGSHUD_20151026_7';
Starting restore at 2015 - 10 - 26 14:23:41
using channel ORA_DISK_1
channel ORA_DISK_1: restoring control file
channel ORA_DISK_1: restore complete, elapsed time: 00:00:01
Finished restore at 2015 - 10 - 26 14:23:42
```

恢复第二个控制文件:

```
RMAN > restore controlfile to '/u01/oradata/yingshudb/control02.ctl' from '/orabak/yingshudbfull_YINGSHUD_20151026_7';
```

有了控制文件,我们就可以把数据库启动到 mount 状态,恢复数据库了。

```
SQL > alter database mount;
Database altered.
```

恢复数据库:

```
RMAN > restore database;
Starting restore at 2015 - 10 - 26 14:30:49
```

```
released channel: ORA_DISK_1
allocated channel: ORA_DISK_1
channel ORA_DISK_1: SID = 10 device type = DISK
channel ORA_DISK_1: starting datafile backup set restore
channel ORA_DISK_1: specifying datafile(s) to restore from backup set
channel ORA_DISK_1: restoring datafile 00001 to /u01/oradata/yingshudb/system01.dbf
... ...
Finished restore at 2015 - 10 - 26 14:34:05
```

还原数据库:

```
RMAN > recover database;
Starting recover at 2015 - 10 - 26 14:37:30
using channel ORA_DISK_1
starting media recovery
archived log for thread 1 with sequence 26 is already on disk as file
... ...
media recovery complete, elapsed time: 00:00:01
Finished recover at 2015 - 10 - 26 14:37:32
```

在 SQLPLUS 下, open 数据库。

```
SQL > alter database open resetlogs;
Database altered.
```

6.4 案例 2 绑定变量问题

核心数据库,分别在 10 月 9 日和 10 月 12 日的 16:50 多分至 17:05 分左右分别出现连接数不够用和大量的"latch: library cache"、"cursor: pin S wait on X"等待事件。下面开始逐步对该等待事件产生原因进行分析。首先,获取该时间段内的 AWR、ADDM 报告进行分析。

1. AWR 报告(见表 6-2)

表 6-2 Top 5 Timed Events(10 月 9 日)

Event	Waits	Time(s)	Avg Wait(ms)	% Total Call Time	Wait Class
latch: library cache	1 542 461	384 954	250	57.9	Concurrency
cursor: pin S wait on X	19 302 183	188 954	10	28.4	Concurrency
CPU time		28 025		4.2	
latch: undo global data	249 410	25 958	104	3.9	Other
latch free	133 284	25 470	191	3.8	Other

2. ADDM 报告

10 月 9 日(16:00—17:00):

```
FINDING 1: 45 % impact (165019 seconds)
--------------------------------
Soft parsing of SQL statements was consuming significant database time.
```

```
    RECOMMENDATION 1: Application Analysis, 45% benefit (165019 seconds)
       ACTION: Investigate application logic to keep open the frequently used
          cursors. Note that cursors are closed by both cursor close calls and
          session disconnects.
    RECOMMENDATION 2: DB Configuration, 45% benefit (165019 seconds)
       ACTION: Consider increasing the maximum number of open cursors a session
          can have by increasing the value of parameter "open_cursors".
       ACTION: Consider increasing the session cursor cache size by increasing
          the value of parameter "session_cached_cursors".
    ------------------------------
Hard parsing SQL statements that encountered parse errors was consuming
significant database time.

    RECOMMENDATION 1: Application Analysis, 31% benefit (114543 seconds)
       ACTION: Investigate application logic to eliminate parse errors.
```

10月12日(16:00—17:00)：

```
Soft parsing of SQL statements was consuming significant database time.
    RECOMMENDATION 1: Application Analysis, 44% benefit (294095 seconds)
       ACTION: Investigate application logic to keep open the frequently used
          cursors. Note that cursors are closed by both cursor close calls and
          session disconnects.
    RECOMMENDATION 2: DB Configuration, 44% benefit (294095 seconds)
       ACTION: Consider increasing the maximum number of open cursors a session
          can have by increasing the value of parameter "open_cursors".
       ACTION: Consider increasing the session cursor cache size by increasing
          the value of parameter "session_cached_cursors".
```

根据 ADDM 的建议，shared pool 解析压力偏大。硬解析量很多，建议把参数 session_cached_cursors 由 20 增大到 500。尽量减轻 shared pool 的解析压力。

hard parse。这里面提到的 SQL 语句均没有使用绑定变量，导致 SQL 语句版本增大，shared pool 压力骤然增大。

建议 open_cursors 增大到 3000。尽量减小出现 open_cursors 不足的错误，但根本还需要应用程序保障游标使用后及时关闭。

通过对比，发现主要还是在异常时段，会话开启的游标数太多（相对于稳定时段）。连接会话的游标开启增大。

10月12日 16:00—17:00 硬解析：(异常)(表6-3)：

表6-3　10月12日 16:00—17:00 硬解析：(异常)

Statistic Name	Time (s)	% of DB Time
parse time elapsed	540 361.71	81.22
hard parse elapsed time	246 266.63	37.02
failed parse elapsed time	214 700.57	32.27
sql execute elapsed time	77 603.29	11.66

下面是给出的建议：

（1）建议所有应用程序使用绑定变量，避免出现 SQL Version Count 过高而导致 shared pool 压力太大的负担。相关 SQL 由下面 SQL 查出。

```
    -- 查需绑定变量的 SQL
col sql for 45
select substr(sql_text,1,40) "sql",
count(*),
sum(executions) "totexecs"
from v$sqlarea
where executions < 5
group by substr(sql_text,1,40)
having count(*) > 30
order by 2;
```

（2）10月9日和10月12日出现两次 ORA-00020：maximum number of processes (1600) exceeded 报错，建议客户把 process 适当增大至 3000。

（3）建议客户把 open_cursors 增大到 3000。把 session_cached_cursors 设置为 500。

（4）对于 ORA-3136 错误，可以按如下方式来进行解决：

```
1：在 listener.ora 文件中添加如下语句：
inbound_connect_timeout_<listener_name> = 0
2：编辑 sqlnet.ora 文件追加如下语句：
sqlnet.inbound_connect_timeout = 0
3：reload listener 使其生效。
 $ lsnrctl reload <listener_name>
```

（5）考虑使用 cursor_sharing=force 参数。

6.5 经典案例 3 数据库无法启动

1．概述

樱澍医院 yszxyy 系统数据库宕机，无法启动，经进一步查询得知是由于存储掉电，主机掉电，操作系统忽然关机所致。数据库在关闭时，alert 没有来得及报任何错。

结论：

（1）因操作系统忽然关机。相当于数据库在不一致的情况下 abort。Buffer 中的数据来不及写到数据文件中去。导致 redo、control file、datafile 的 SCN 号不同步。在启动的时候起不来，报 ORA-600 错误。

（2）数据库没有备份。既没有 RMAN 全备，也没有近期的 exp 导出备份。

（3）local_listener 参数没有设置，连接不稳定，时断时续。

解决此问题的方法如下：

（1）加隐含参数在不一致的情况下强制启动数据库。

（2）重建 undo 表空间，将旧的 undo 表空间删除。

(3) 将数据导出。
(4) 重建新数据库。
(5) 将数据导入新数据库,旧库作废。

针对本经典案例的主要建议如下:

在本次故障处理中发现了一些问题,具体描述和建议会在下面的报告中详细阐述。

表 6-4 是一些主要问题和建议的总结。

表 6-4 一些主要问题和建议

No.	主要问题	说明及参考	建议解决时间
1	定期删除 alert.log 及 trace 文件	Alter 日志过大,vi 无法打开。本库的 alter 日志有 600MB	近期解决
2	定期删除 listener.log	Listener.log 有 2GB 大	近期解决
3	硬件已过保,建议新置硬件	硬件已经运行 7 年以上,建议更新换代	近期解决
4	注意磁盘空间利用率	使用 df -h	近期解决
5	定期进行 RMAN 全备	便于恢复	近期解决
6	建议实现数据库的高可用,如 DataGuard。	建议建立灾备系统	近期解决

我们在本次检查过程中,已就大多数问题及建议与樱澍医院的 DBA 进行了交流,并详细告之了具体的实施方法。

2. 现场处理

(1) 检查数据库 alert 日志。查看警告日志在什么地方:

```
sqlplus /nolog
conn /as sysdba
show parameter dump
```

警告日志没报错:

```
Thread 1 advanced to log sequence 1331 (LGWR switch)
  Current log # 2 seq # 1331 mem # 0: + ORACLEASMDS3400/yszxyy/onlinelog/group_2.258.702315483
  Current log # 2 seq # 1331 mem # 1: + ARCHDG/yszxyy/onlinelog/group_2.258.702315485
Mon Aug 2 07:48:52 2015
Starting ORACLE instance (normal)
LICENSE_MAX_SESSION = 0
```

(2) 强制启动数据库:

```
SQL> conn /as sysdba
Connected to an idle instance.
SQL> startup nomount pfile = '/tempfs/pfile.ora';
ORACLE instance started.
```

```
Total System Global Area 2516582400 bytes
Fixed Size                  2086032 bytes
Variable Size             905972592 bytes
Database Buffers         1593835520 bytes
Redo Buffers               14688256 bytes
SQL> alter database mount;

Database altered.
SQL> recover database until cancel;
ORA-00279: change 240795288 generated at 08/03/2015 00:16:24 needed for thread 1
ORA-00289: suggestion : +ARCHDG
ORA-00280: change 240795288 for thread 1 is in sequence #3
Specify log: {<RET>=suggested | filename | AUTO | CANCEL}
cancel
ORA-01547: warning: RECOVER succeeded but OPEN RESETLOGS would get error below
ORA-01194: file 1 needs more recovery to be consistent
ORA-01110: data file 1: '+ORACLEASMDS3400/yszxyy/datafile/system.260.702315491'


ORA-01112: media recovery not started
startup mountpfile = '/tempfs/pfile.ora';
ALTER SESSION SET EVENTS '10015 TRACE NAME ADJUST_SCN LEVEL 3';
alter database open;
Database altered.
```

(3) 建立新的 undo 表空间：

```
create undo tablespace undotbs3 datafile '+ORACLEASMDS3400/yszxyy/datafile/undotbs3'
size 1000m;
alter system set undo_tablespace='UNDOTBS3' scope=spfile;
shutdown abort;
startup up;
drop tablespace undotbs1;
```

(4) 将数据库导出：

```
exp 'sys/oracle as sysdba' buffer=20971520 full=y grants=y file=/tempfs/fulldb.dmp log=/tempfs/impfull_100803.log
```

(5) 新建数据库，使用 dbca 建库：

```
dbca
```

(6) 将数据库整库导入：

```
imp sys/oracle buffer=20971520 full=y grants=y ignore=y file=/tempfs/fulldb.dmp log=/tempfs/impfull_100803.log
```

(7) RMAN 备份：省略。参见第 5 章，制订合适的备份策略。

(8) 设置参数 local_listener,解决连接问题:

```
主机 12
alter system set local_listener = '(address = (protocol = tcp)(host = 192.168.0.10)(port = 1521))' scope = both sid = 'YSZXYY1';
alter system set local_listener = '(address = (protocol = tcp)(host = 192.168.0.11)(port = 1521))' scope = bothsid = 'YSZXYY2';
```

6.6 案例 4 OEM bug

樱澍 ysiis 系统/home/oracle/出现异常的三种文件,heapdump、javacore、snap0001,经进一步诊断确认是由于数据库 oem bug 所致。经过和客户商议,暂时关闭 oem 功能,观察以后的状况。若不再出现此类问题,可以永久禁用 oem,直至升级到 11.2.0.4 再启用。问题解决。

通过案例 4 可以得到以下结论:

Oracle 11g,11.1.7 不是 11g 最稳定的版本,有 bug 比较难免。也可以考虑升级,升级前做好测试工作。

解决此问题的方法如下:

(1) 停止 oem Web 服务。
(2) 停止 oem 服务。

```
emctl stop dbconsole
emctl stop agent
```

6.7 案例 5 网络故障

应客户要求,于 2015-6-29 上午 11 点到达现场,对 ora-29740 错误进行分析,如表 6-5 所示。

表 6-5 ora-29740 错误分析

问题描述	内容	处理结果	潜在影响/后续措施
Trace dumping is performing id =[cdmp_20150629192723] MonJun 29 19:31:43 2015 Errors in file /app/oracle/admin/ ysccsm/bdump/ysccsm2_lmon_ 19538.trc: ORA-29740: evicted by member 1, group incarnation 52	MonJun 29 19:31:43 2015,RAC 数据库 ysccsm2 实例,发生宕机	事故原因: 私有网络通信故障。 RAC 负载过重。 是否已经考虑 RAC 节点分开应用,避免交叉访问	建议布署 OSW。 监控网络 priv 网卡通信情况。 监控 Oracle 数据库压力情况

1. 现场处理

1) alert_ysccsm2.log

```
Mon Jun 29 19:27:27 2015
Trace dumping is performing id=[cdmp_20150628192723]
Mon Jun 29 19:31:43 2015
Errors in file /app/oracle/admin/ysccsm/bdump/ysccsm2_lmon_19538.trc:
ORA-29740: evicted by member 1, group incarnation 52
Mon Jun 29 19:31:43 2015
LMON: terminating instance due to error 29740
Mon Jun 29 19:31:43 2015
Errors in file /app/oracle/admin/ysccsm/bdump/ysccsm2_lms0_19542.trc:
ORA-29740: evicted by member , group incarnation
Mon Jun 29 19:31:43 2015
Errors in file /app/oracle/admin/ysccsm/bdump/ysccsm2_lms1_19544.trc:
……
Tue Jun 29 09:34:17 2015
Starting ORACLE instance (normal)
```

2) ysccsm2_lmon_19538.trc

```
*** 2015-06-29 19:19:57.115
kjxgrcomerr: Communications reconfig: instance 0 (51,51)
Submitting asynchronized dump request [2]
kjxgrrcfgchk: Initiating reconfig, reason 3
*** 2015-06-29 19:20:02.185
kjxgmrcfg: Reconfiguration started, reason 3
kjxgmcs: Setting state to 51 0.
*** 2015-06-29 19:20:02.190
      Name Service frozen
kjxgmcs: Setting state to 51 1.
*** 2015-06-29 19:20:02.484
Obtained RR update lock for sequence 51, RR seq 51
```

3) alert_ysccsm1.log

```
Mon Jun 29 19:19:57 2015
Trace dumping is performing id=[00628191957]
Mon Jun 29 19:21:43 2015
Waiting for clusterware split-brain resolution
Mon Jun 29 19:26:06 2015
Errors in file /app/oracle/admin/ysccsm/bdump/ysccsm1_diag_7589.trc:
ORA-00600: internal error code, arguments: [17182], [0x1037DF9C8], [], [], [], [], [], []
Mon Jun 29 19:27:15 2015
```

```
Restarting dead background process DIAG
DIAG started with pid = 3
```

4) alert_ysccsm1.log

```
*******************************************
HEAP DUMP heap name = "pga heap" desc = 1036a6030
 extent sz = 0x2190 alt = 168 het = 32767 rec = 0 flg = 3 opc = 2
 parent = 0 owner = 0 nex = 0 xsz = 0x4068
EXTENT 0 addr = 107d5bc98
  Chunk        107d5bca8 sz =      16472     freeable "KST Heap      " ds = 1036ab250
… …
6C271444:00000008      3      0 10401    28 KSXPMAP: client 1 base 0x3800c0000 size 0x8d6f40000
KSTDUMP: End of in-memory trace dump
DIAG detachs from CM
error 600 detected in background process
OPIRIP: Uncaught error 447. Error stack:
ORA-00447: fatal error in background process
ORA-00600: internal error code, arguments: [17182], [0x1037DF9C8], [], [], [], [], [], []
```

2. Ora-29740 原因分析

可能出现 ora-29740 故障的原因：

```
Common causes for an ORA-29740 eviction (Reason 3):

   a) Network Problems.
   b) Resource Starvation (CPU, I/O, etc..)
   c) Severe Contention in Database.
   d) An Oracle bug.

Tips for tuning inter-instance performance can be found in the following note:

 Note 181489.1
 Tuning Inter-Instance Performance in RAC and OPS
```

排查问题需查看的重要文件：

a) Each instance's alert log

b) Each instance's LMON trace file

c) each instance's LMD and LMS trace files

d) Statspack reports from all nodes leading up to the eviction

e) Other bdump or udump files…

f) Each node's syslog or messages file

g) iostat output before, after, and during evictions

h) vmstat output before, after, and during evictions

i) netstat output before, after, and during evictions

There is a tool called "OS Watcher" that is being developed that helps gather this information. For more information on "OS Watcher" see Note 301137.1 "OS Watcher User Guide".

因故障当时压力情况,通信情况无法重现,只能根据现有的资料做出判断。

下面列出出现 ora-29740 故障的原因:

(1) 私有网络通信故障。

(2) RAC 负载过重,建议监控 CPU 利用率,心跳线压力情况。

3. Ora-600 原因分析

按时间顺序列举如下:

(1)

```
*** 2015 - 06 - 29 19:19:57.115
kjxgrcomerr: Communications reconfig: instance 0 (51,51)
Submitting asynchronized dump request [2]
kjxgrrcfgchk: Initiating reconfig, reason 3
*** 2015 - 06 - 29 19:20:02.185
kjxgmrcfg: Reconfiguration started, reason 3
```

19 点 19 分 57 秒数据库出现通信故障。

(2)

```
Mon Jun 29 19:21:43 2015
Waiting for clusterware split - brain resolution
```

19 点 21 分 43 秒,等待集群剔除实例。

(3)

```
Mon Jun 29 19:26:06 2015
Errors in file /app/oracle/admin/ysccsm/bdump/ysccsm1_diag_7589.trc:
ORA - 00600: internal error code, arguments: [17182], [0x1037DF9C8], [], [], [], [], [], []
```

19 点 26 分 06 秒。因通信故障,集群决定剔除实例,报出 ora-600 错误。

(4)

```
Mon Jun 29 19:31:43 2015
Errors in file /app/oracle/admin/ysccsm/bdump/ysccsm2_lmon_19538.trc:
ORA - 29740: evicted by member 1, group incarnation 52
```

19 点 31 分 43 秒。报 ora-29740 错误,实例 ysccsm2 被剔除。

ora-600 的起因是由于通信故障,集群决定剔除实例时报出的,和 Oracle bug 无关。

4. 处理方法

(1) 用 OSW 工具监控私有网卡通信情况。

(2) 监控 Oracle 数据库压力、CPU 情况等。开启 statspack 收集,查看 statspack 报告,分析 Oracle 压力情况。

5. 处理结果

OSW 已部署。结果显示,两张私有网卡在长时间 traceroute 时,发生间歇性中断。

6.8 案例 6 数据库版本等问题

1. 背景

ysms 系统承担着公司的核心业务,为全省用户提供服务,该系统运行在 AIX 5.3 上的 Oracle(10.2.0.1)。在最近的运行过程中,数据库系统的性能问题比较严重,会导致前台响应变慢,业务处理时间变长,CPU 瓶颈比较突出,利用率长时间持续 90% 以上。

针对这些问题,我们对 ysms 数据库系统进行首次的优化和故障排除,优化方面包括数据库参数、CPU。不涉及其他方面的调整。为了更好地完成对 ysms 数据库系统的优化,我们对 ysms 系统进行多次的性能评估,收集了相关的信息,本着安全第一的原则,对参数做出基础的调整。

2. 性能优化的定义和范围

- CPU 时间;
- CPU 等待时间(wait time);
- 系统 CPU(usr+sys%);
- 数据库参数。

6.8.1 数据库 ysms 的首次优化方案

1. 参数配置的调整说明

(1) 参数调整 processes

```
alter system set processes = 500            scope = spfile;
```

说明:此参数影响允许连接进入数据库的连接数。500 为一般设置。已成功设置。

(2) 参数调整 open_cursors

```
alter system set open_cursors = 1000        scope = both;
```

说明:此参数是数据库可用的游标的个数,1000 是一般设置。已成功设置。

(3) 参数调整 sga

```
alter system set sga_target = 4294967296    scope = spfile;
alter system set sga_max_size = 4294967296  scope = spfile;
```

说明:由 AWR 报告可以看出,Shared Pool Size:416MB,明显偏小。从 alter 里也可以看出此问题存在。Memory Notification:Library Cache Object loaded into SGA。

加大 SGA 只是保守的调整,调整后观察其运行情况,如还不达标,最后调整将使用手工管理 SGA 为最佳。

已成功设置手工管理 SGA。

2. CPU 利用率的调整说明

通过 topas 可以看出,CPU 利用率使用过高。调整之前利用率长时间持续在 95% 左右。其中有两到三个进程,占据了大概一半 CPU 资源。一个进程占一颗 CPU,并且不释放资源一直在占用。真正留给应用的 CPU 只有 50%。调整占用资源的两个进程,是 CPU 调优的基础所在。以下是关于调整 top 进程,让其释放占用的 CPU 资源的解决方案。

3. 数据库 ysms 优化实施步骤

(1) 停掉占用 CPU 的 job

通过 AWR 报告可以看出,其中的 job 占用大量的 CPU time。停掉应用之后,确定这个运行的 job 是 182。此 job 调用的存储过程是 BILL.p_dailyliwang,此存储过程中有一条 SQL,消耗了一颗 CPU。

```
Elapsed      CPU                         Elap per    % Total
Time (s)    Time (s)    Executions      Exec (s)    DB Time    SQL Id
--------    --------    ----------      --------    -------    ------
  3,510      2,741           0            N/A         21.5     b10bwp8vtnytz
INSERT INTO DAILYLIWANGDETAIL_TABLE SELECT 4, 1, CITYCODE, USERID, USERNAME, LOG
INNAME, BINDINGPHONE, BRANCH, LOCKTYPE, CONNECTTYPE, BEGINTIME, RATE, PREPAID, S
TATE, :B2 FROM ( SELECT A.USERID USERIDX, (SELECT CITYCODE FROM TMP_DAILYLIWANGD
ETAIL_TABLE B WHERE A.USERID = B.USERID AND B.MONTHS = :B2 AND ROWNUM = 1) CITYCODE, (
```

通过应用工程师的确认,此 job 可以停止,不影响业务。所以停止此 job,便可以释放 CPU。

```
exec dbms_job.remove(182).
```

停掉此 job 后,CPU 资源已成功释放。

(2) 停掉 Oracle 自身进行统计信息收集的 SCHEDULER

```
exec DBMS_SCHEDULER.DISABLE('GATHER_STATS_JOB');
```

在数据发生大量变化之后,如表里的数据变化超过 10%,可以手工收集统计信息。

停掉此 SCHEDULER 后,资源已成功释放。

4. 调整前后的对比

CPU 利用率由原来的 95% 降至现在的平均 45% 左右,释放了近一半的 CPU 资源。响应时间也相应提高。由于时间的原因,具体的响应时间对比就不一一列出了。

5. 下一步的工作重点

表 6-6 为下一步的工作重点。

表 6-6 下一步的工作重点

No.	问题描述	解决时间
1	数据库版本需要升级。割接至 570 之后,可以解决此问题。升级至 10.2.5.12 或 11.2.0.4.7	重要。马上解决
2	某些索引的碎片需要整理	近期解决
3	SQL 调整,可以针对某个业务模块的 SQL 进行优化分析	近期解决

续表

No.	问 题 描 述	解 决 时 间
4	数据库参数,操作系统参数,进一步调研,调整	长期解决
5	数据库 shared_pool,soft parse 指标不能满足要求,SQL 语句存在没有使用绑定变量的情况	较重要,长期解决
6	确认备份的有效性	重要,马上解决

6.8.2 数据库 ysms 的第二次优化方案

1. 背景

ysms 数据库运行在 Oracle 10 上,具体版本为 10.2.0.1 on AIX。该版本 10.2.0.1 是 Oracle 10g 的一个最初发行的版本,版本较低,不是 Oracle 10g 中最稳定的版本,由于该版本存在的 bug 导致数据库出现了多个 ora-错误。10.2.0.5.12 是 Oracle 10g 的最终版本,不仅可以解决前面提到的 bug 问题,10.2.0.5.12 本身稳定性要远高于 10.2.0.1。由于系统要扩容,本系统需切换主机。存储扩容,但不更换存储,本文档对 Oracle 割接方案及从 10.2.0.1 升级到 10.2.0.5.12 的方案进行了描述。

2. 升级实施方案

由于目前系统为生产系统,我们不仅需要保证升级及割接的成功,还需要保证升级及割接给业务带来的影响最小,需要在对生产环境进行升级割接前进行测试。

升级割接测试有三个目的,一是检查 Oracle 升级过程中会遇到哪些问题,这些问题应该怎么解决。二是掌握升级的每一步大概需要花费多少时间。三是测试应用在新的版本上是否会出现问题,出现的问题是否可以解决,如何解决。

方案一:导出导入方式,测试过程

(1) 用 exp,imp 方式。这种方式最安全。但时间窗口需测试。实际割接过程需停数据库。但测试过程中无须停数据库。这种方式的好处是将整个数据库的数据整理了一遍,对性能将有很好的帮助。

(2) 在测试机上安装 10.2.0.5 软件,详见<<Oracle_db_安装配置_付培利_2015.11.5.doc>>。

(3) 在测试机上创建数据库。注意和生产机上的数据库名及数据文件目录名等要一致、空间够用。

(4) 全库方式导出整个生产库。

全库方式,将数据库中的所有对象导出(根据需要更改路径、名称等):

```
exp user/pwd file = /dir/xxx.dmp log = xxx.log full = y commit = y buffer = 50000000
```

(5) 全库方式导入整个测试库。注意一定要记 log,导入完成后,仔细查看 log,一些错误可以忽略。查看是否有不可以忽略的错误。导入脚本:

```
imp user/pwd file = /dir/xxx.dmp log = xxx.log ignore = y commit = y grants = y full = y
```

(6) 导入成功后,通知应用部门,进行业务测试,功能性测试。

(7) 排查数据库出现的异常,兼警告日志等。
(8) 测试成功。

方案二:冷备方式测试

(1) 准备测试环境,在测试机上安装 10.2.0.5 软件。详见<<Oracle_db_安装配置_付培利_2015.11.5.doc>>。
(2) 在测试机上准备和生产库对应目录的存储空间。
(3) 停止生产库:

```
lsnrctl stop
shutdown immediate;
```

(4) 将生产系统上的数据库复制到测试机上;cp datafile,记录过程中所花的时间。
需 cp 的文件有 spfile、密码文件、控制文件、redo.log、数据文件。
(5) 在测试机上建立对应的 dump 目录。
(6) 按后面描述的升级过程对测试环境 Oracle 数据库升级(10.2.0.1 到 10.2.0.5),记录过程中出现的问题和解决方法,记录升级数据库所花的时间。
(7) 对测试环境数据字典进行升级,记录过程中出现的问题和解决方法,记录升级数据字典所花的时间。
(8) 在测试库上创建 listener。使用 netca 工具启动 listener。
(9) 升级数据字典;以下描述了升级步骤,但仍然建议在实际操作过程中参考下面的文档。

- Oracle® Database Patch Set Notes
- 10g Release 2 (10.2.0.5) Patch Set 4 for AIX
- Part Number E15233-02
- Dbua 图形界面的方法。

或者采用手工升级的方式。预升级检查。

```
sqlplus '/ as sysdba'
startup upgrade;
spool upgrade_info.log
@?/rdbms/admin/utlu102i.sql
spool off
select * from dba_registry;
```

检查 upgrade_info.log,确认没有问题
对数据库进行升级:

```
shutdown immediate
startup upgrade
spool patch.log
@?/rdbms/admin/catupgrd.sql
Spool off
Shutdown immediate;
```

```
Startup;
@?/rdbms/admin/utlrp.sql
select comp_name, version, status from sys.dba_registry;
```

(10) 在升级后的测试环境中对应用进行测试,测试应用在新的版本上是否会出现问题,出现的问题是否可以解决,如何解决。

(11) 最终采用方案二。

3. 回退方案

一旦升级失败,就需要回退。回退的方法就是恢复至没升级前的状态。

因为我们的生产库(旧库)本身没有做任何操作,只是把数据库停下来。也没有动原来的数据库文件,只是 cp 了一份。如果在测试(新环境)中升级失败,我们只需要启动旧库即可,无须恢复测试库。如遇需恢复新环境的情况。具体方法如下:

(1) 在新节点上恢复备份的 Oracle 软件(10.2.0.1);

(2) 重跑 10.2.0.1 的 catalog 脚本;

(3) 如第(2)步不成功,恢复数据库使用 imp 方式导入整库,这步时间较长。

6.9 案例 7 数据库改造方案(简版)

1. 数据库系统的现状及故障分析

目前数据库服务器的操作系统为 Windows,数据库版本为 10.2.0.1RAC。Windows 操作系统运行 RAC 及现有的管理方式有以下不足:

(1) 多次出现数据库 down 机,导致应用系统不能正常提供服务。

(2) Windows+Oracle RAC 方式很不稳定。目前运行在 Windows 上的 RAC 数据库服务器不超过 10%。

(3) 数据库管理不正规,建了三个 RAC 数据库,一个单机版数据库,占用大量系统资源,相互影响导致操作系统宕机。

(4) 数据库版本为 10.2.0.1,是 Oracle 发行的基础版本,没有安装任何补丁,很容易触发 Oracle bug,存在严重不稳定因素。

(5) 樱澍 DBA 现场处理过两次宕机事故。一次是因为数据库内存占用过大,导致系统宕机。第二次因为数据库参数设置不正确导致停机。

2. 改造升级后的优势介绍

这次系统改造之后,彻底解决了以上问题。主要描述如下:

(1) 硬件环境得到了提升,新服务器配备:32CPU,128GB 内存,系统资源充分。

(2) 软件环境全新改版,操作系统采用 Redhat6.6 操作系统。Linux 操作系统比 Windows 服务器操作系统更加稳定。

(3) 数据库采用 Oracle 11g 最稳定的版本 11.2.0.4.7。数据库版本的升级将使数据库运行的效率大大增加,避免了 bug 的触发。

(4) 存储划分采用性能最优的 RAID10 作为数据库主要的存储地,部分对效能要求不高的可以采用 RAID5 存储数据。

(5) 备份采用数据库安全的备份方式 RMAN 备份。可以实现数据库的全恢复,保证数据库安全。

（6）账号管理，运维 DBA 统一管理数据库账号，密码设计不可破解。改变了目前多人拥用 root 和数据库 DBA 账号密码的现状，使数据库更加安全。

（7）数据集中使原有的数据库有 4 个数据库合为 1 个数据库，充分利用了系统资源，降低了管理和运行维护成本，使数据库便于管理，内存资源可以得到良好的利用，在今后的排错中，更易于找出问题的错误点。如果今后有分拆数据库的需求，可以根据用户灵活地拆成两个数据库。

3．实施步骤及需要甲方配合的内容（见表 6-7）

表 6-7 实施步骤及需要甲方配合的内容

序号	环境准备内容	负责人	状态	备注
1	检查操作系统环境是否满足 Oracle 安装需求	樱澍 DBA	完成	9月1日
2	安装 Oracle crs 软件 11.2.0.4	樱澍 DBA	完成	9月1日
3	安装 Oracle 数据库软件 11.2.0.4	樱澍 DBA	完成	9月1日
4	升级 Oracle crs 至 11.2.0.4.7	樱澍 DBA	完成	9月1日
5	升级 Oracle 数据库软件至 11.2.0.4.7	樱澍 DBA	完成	9月1日
6	创建数据库 ysdb	樱澍 DBA	完成	9月3日
7	根据软件开发商需求建应用表空间	樱澍 DBA	完成	软件商确认正式应用的表空间
8	建 Oracle 数据库用户	樱澍 DBA	完成	9月3日
9	跟踪数据库的稳定性	樱澍 DBA	完成	9月4日
10	准备测试数据库	软件商	完成	9月14日-15日
11	导入新建的数据库进行测试	软件商	完成	9月16日
12	功能及压力测试	软件商	完成	9月19日-21日
13	删除测试数据用户	软件商	完成	9月22日
14	测试情况总结及数据库迁移准备会议	三方	完成	9月22日
15	停止所有业务	甲方	完成	9月23日 18:00
16	导出正式数据库	软件商	完成	9月23日 18:30
17	导入新的正式数据库环境	软件商	完成	9月24日-25日
18	切换正式数据库环境网络	网络	完成	
19	跟踪数据库的稳定性	樱澍 DBA	完成	
20	试运行	甲方	完成	项目成员值班

6.10 案例 8 数据库参数设置问题

应客户要求，于 2013-11-03 日到达现场，对 YSEIP 数据库迁移后出现的性能问题进行分析处理，如表 6-8 所示。

第6章 数据库故障处理

表6-8 对YSEIP数据库迁移后出现的性能问题进行分析处理

问题描述	内 容	处理结果	潜在影响/后续措施
YSEIP系统数据库由原Sun主机迁移到新IBM主机后,系统性能极差	现场调查问题发生的现象及其他状态,检查操作系统版本、数据库版本及补丁、alert.log等,监控系统资源,检查数据库参数配置等,对问题进行分析,并提出解决方案	找到问题原因,提出解决方案,实施了补丁安装	系统监控,优化应用

1. 现状及检查

(1) 系统配置:

两节点 Oracle9.2.0.7 RAC on AIX 5300-05-CSP,HACMP 5.3,每节点4CPU,7.7GB内存,9.2.0.7版未安装 ONE-OFF PATCH。

(2) 现象:

Page in/out 多,IO 大,CPU 空闲为0,应用响应慢。

(3) 系统变化:

原来三个数据库(ysdb、orcl、yssm)实例跑在一台Sun服务器上,为了系统升级,暂时把它们迁移到一对IBM p550的集群上,把ysdb配置为两节点RAC,orcl跑在节点1上,yssm跑在节点2上。

原来的Sun服务器为8CPU、16GB内存,新的IBM服务器每节点只有4CPU、7.7GB内存。

Oracle版本由原9.2.0.5 for Solaris 变为9.2.0.7 RAC for AIX5L。

2. 原因分析

造成系统性能严重下降的原因有以下几点:

(1) 新服务器内存只有7.7GB,为原系统的一半,而数据库的初始化参数没有相应调整,ysdb的SGA为4GB,orcl和yssm的SGA分别在1.7GB左右,这样,在每个节点上SGA占用了将近6GB内存,造成物理内存严重不足,导致大量page in / page out,IO量加大,CPU消耗加大。

(2) Oracle 9.2.0.7 存在 bug,使得 ksu process alloc latch yield 等待消耗大量CPU资源,需要安装补丁 p4947798_92070_AIX64-5L.zip 来解决,也可以升级到9.2.0.8版本。

(3) 新服务器CPU数量少于原服务器。

3. 处理方法

(1) 优化全部4个实例的初始化参数,减小不必要的 java_pool_size、large_pool_size 等。

通过以下调整,每节点的上两个实例的SGA总和控制在3.5GB以内。

Ysdb,如果 orcl 和 yssm 不迁出
*.db_cache_size = 2306867200
*.java_pool_size = 134217728
*.large_pool_size = 67108864
*.pga_aggregate_target = 2147483648
*.sga_max_size = 4194304000
*.shared_pool_size = 314572800
=>
*.db_cache_size = 1.5g
*.java_pool_size = 20m
*.large_pool_size = 20m
*.pga_aggregate_target = 1g
*.sga_max_size = 不设,或设到 2.4g

Ysdb,如果 orcl 和 yssm 迁出
*.db_cache_size = 2306867200
*.java_pool_size = 134217728
*.large_pool_size = 67108864
*.pga_aggregate_target = 2147483648
*.sga_max_size = 4194304000
*.shared_pool_size = 314572800
=>
*.db_cache_size = 2g
*.java_pool_size = 20m
*.large_pool_size = 50m
*.pga_aggregate_target = 1g
*.shared_pool_size = 400m
*.sga_max_size = 3.2g 或 3g

orcl
*.java_pool_size = 536870912
*.large_pool_size = 117440512
*.sga_max_size = 1797231144
*.shared_pool_size = 536870912
=>
*.java_pool_size = 20m
*.large_pool_size = 20m
*.sga_max_size = 不设或设为 1g
*.shared_pool_size = 300m

yssm
*.db_cache_size = 536870912
*.java_pool_size = 83886080
*.large_pool_size = 117440512
*.pga_aggregate_target = 251658240
*.processes = 150
*.sga_max_size = 1411354624
*.shared_pool_size = 637534208
=>

*.db_cache_size = 536870912

*.java_pool_size = 20m

*.large_pool_size = 2m

*.pga_aggregate_target = 251658240

*.processes = 150

*.sga_max_size = 不设或设为 1g

*.shared_pool_size = 300m

(2) 安装以下 Oracle 补丁：

p4947798_92070_AIX64-5L.zip 可以大大减轻 CPU 消耗

p5496862_92070_AIX64-5L.zip 操作系统为 AIX5300-05 时必打的 IO 相关的补丁

其他可选的补丁(这些补丁与本次性能问题无关,在 9.2.0.7 上遇到的可能性较大,建议安装):

p4925103_92070_AIX64-5L.zip WRONG RESULT OCCURS BY USING MAX() FOR NULL DATA AFTER UPGRADED 9.2.0.7

p4192148_92070_AIX64-5L.zip VIEW DEFINITION IN DBA_VIEWS SHOWS WRONG SYNTAX

p4721492_92070_AIX64-5L.zip INDEX REBUILD GETS NO ORA-54 WHEN IT WAITS FOR A DML

注:安装补丁需要使用 OPatch 工具,该工具以一个补丁的形式提供,补丁号为 2617419.

(3) 条件允许的话,增加 CPU 和物理内存。

(4) 调整其他参数(与本次性能问题无直接关系,但建议调整):

相关网络参数的当前设置:

ipqmaxlen = 100

sb_max = 1310720

udp_sendspace = 65536

udp_recvspace = 262144

rfc1323 = 1

tcp_sendspace = 262144

tcp_recvspace = 262144

以上黑体的两个参数偏小,需要调大:

no -r -o ipqmaxlen = 512

no -p -o udp_recvspace = 655360

虚存管理相关参数当前设置:

minperm% = 20

maxperm% = 80

maxclient% = 80

建议调整:

vmo -p -o minperm% = 10

vmo -p -o maxclient% = 20

vmo -p -o maxperm% = 20

```
oracle 用户的 ulimit 设置:
default:
        fsize = -1
        core = 2097151
        cpu = -1
        data = 262144
        rss = 65536
        stack = 65536
        nofiles = 2000


建议增加以下行:
oracle:
        fsize = -1
        cpu = -1
        data = -1
        rss = -1
        stack = -1
```

(5) 监控并优化应用:

在上述方案实施后,持续监控系统运行,发现消耗资源较大的 SQL 语句后及时进行优化,进一步减小对系统资源的消耗。

4. 处理结果

补丁已安装,安装过程中发现节点 2 的 inventory 有问题无法安装,从节点 1 复制过来。

```
p4947798_92070_AIX64-5L.zip 可以大大减轻 CPU 消耗
p5496862_92070_AIX64-5L.zip 操作系统为 AIX5300-05 时必打的 IO 相关的补丁
```

OS 及 Oracle 参数已调整,orcl 和 yssm 已决定迁移到其他机器。

5. 其他处理

(1) 发现 v$thread 视图中显示 thread 2 为 PRIVATE,是手工建立第二实例时的问题,已修改为 PUBLIC。

(2) 建立了 statspack 的 crontab,每半小时收集信息用于优化。

(3) 建立了 crontab,每天零点对 ysbm 和 ysbm 用户进行表分析(原来使用的 Oracle job 已失效),后根据开发商要求,增加了对 YSEIPORADB 用户进行表分析。

```
0,30 * 3,4,5 12 * /home/oracle/statspack/statspack_snap.sh >/dev/null 2>&1
0 0 * * * /home/oracle/statspack/gather_stats.sh >/dev/null 2>&1
```

```
#!/usr/bin/ksh
ORACLE_BASE=/u01/app/oracle; export ORACLE_BASE
ORACLE_HOME=$ORACLE_BASE/product/9.2.0; export ORACLE_HOME
ORACLE_SID=ysdb1; export ORACLE_SID
ORA_NLS33=$ORACLE_HOME/ocommon/nls/admin/data; export ORA_NLS33
LIBPATH=$ORACLE_HOME/lib:$LD_LIBRARY_PATH; export LIBPATH
PATH=$ORACLE_HOME/bin:/usr/ccs/bin:/usr/opt/networker/bin:$PATH; export PATH
NLS_LANG=AMERICAN_AMERICA.ZHS16GBK; export NLS_LANG
```

```
NLS_TERRITORY = ; export NLS_TERRITORY
NLS_DATE_FORMAT = ; export NLS_DATE_FORMAT
NLS_DATE_LANGUAGE = ; export NLS_DATE_LANGUAGE

sqlplus -S /nolog << EOF
connect / as sysdba
exec dbms_stats.gather_schema_stats(ownname => 'BPM47', cascade => TRUE);
exec dbms_stats.gather_schema_stats(ownname => 'BPM49', cascade => TRUE);
exec dbms_stats.gather_schema_stats(ownname => 'YSEIPORADB', cascade => TRUE);
exit
EOF
```

```
#!/usr/bin/ksh
ORACLE_BASE = /u01/app/oracle; export ORACLE_BASE
ORACLE_HOME = $ ORACLE_BASE/product/9.2.0; export ORACLE_HOME
ORACLE_SID = ysdb1; export ORACLE_SID
ORA_NLS33 = $ ORACLE_HOME/ocommon/nls/admin/data; export ORA_NLS33
LIBPATH = $ ORACLE_HOME/lib: $ LD_LIBRARY_PATH; export LIBPATH
PATH = $ ORACLE_HOME/bin:/usr/ccs/bin:/usr/opt/networker/bin: $ PATH; export PATH
NLS_LANG = AMERICAN_AMERICA.ZHS16GBK; export NLS_LANG
NLS_TERRITORY = ; export NLS_TERRITORY
NLS_DATE_FORMAT = ; export NLS_DATE_FORMAT
NLS_DATE_LANGUAGE = ; export NLS_DATE_LANGUAGE

sqlplus -S /nolog << EOF
connect perfstat/perfstat
exec statspack.snap(i_snap_level => 7);
exit
EOF
```

6.11　案例 9　回闪区的限额被占满

1. 背景

2015 年 1 月 20 日，应北京樱澍公司的要求，Oracle 数据库工程师于 2015 年 1 月 20 日赶赴现场 standby，对期间出现的问题进行了详细诊断。

2. 问题处理情况

针对北京樱澍公司数据库系统出现的问题，Oracle 工程师详细进行了检查，并实施相应的调整。现对该处理过程做一个总结，请参看表 6-9。

3. 问题处理总结

目前北京樱澍公司数据库系统运行正常，问题得到了解决。一些后续问题还待继续监控，我们建议客户进一步跟踪问题。如需在数据库性能方面的详细检查，请继续选择数据库性能检查。

目前数据库的版本是 11.2.0.1，建议升级至 11.2.0.5。

表 6-9　问题处理总结

问题描述	内　　容	解决结果	潜在影响及后续措施计划
数据库启动错误 ORA-3113 "end of file on communication channel	数据库无法启动 查看 trace 文件发现： ORA-19815： WARNING： db_recovery_file_dest_size of 50000m	将回闪区大小改为 80GB。/已解决	
归档日志无法正常自动删除	crosscheck archivelog all;	手工删除归档日志/已解决	

本次服务详细过程如下。

(1) 背景概述

数据库启动错误：

```
ORA - 3113 "end of file on communication channel
```

单从警告日志无法判断问题的根源。进一步查询 trace 文件。发现：

```
ORA - 19815:
WARNING:
db_recovery_file_dest_size of 50000m
```

(2) 现场处理

故障/问题 #1

故障描述：

数据库无法启动，报错 ora-03113

原因分析：

归档日志已经将回闪区占满。

建议将 db_recovery_file_dest_size 更改至 80GB。

建议将归档日志删除两天之前的。

解决措施：

```
alter system set db_recovery_file_dest_size = 80000m scope = both sid = ' * ';
重新启动数据库,两边都需要重启.
crosscheck archivelog all;
delete noprompt archivelog until time '(SYSDATE - 1)';
```

处理结果：

已解决。

第 7 章 数据库调优艺术

数据库的性能至关重要。在速度慢的时候,必须有性能优化的意识,不能一味地增加硬件。128GB 内存,128 颗 CPU 不一定使得上。常常见到 96GB 内存,使用了 6GB,只用了一个零头,白白浪费了硬件,硬件不会自动用上,需要我们调整。

曾经见过一个客户,由于性能慢,使用系统的客户端部门和管理运维的部门吵的不可开交。花点钱做个调优不就完了嘛,由于甲方流程原因根本走不动调优这条路,我只能感到惋惜和遗憾了。这么好的机器,这么慢的速度,可惜。

7.1 性能问题存在的背景

随着单位业务的发展,数据库的数据量呈线性增长。系统所承受的压力越来越大。运行几个月或几年,数据库系统很容易出现性能问题,可能会导致前台响应变慢,业务处理时间变长,CPU 瓶颈突出,内存不足,I/O 瓶颈等问题。特别是在业务高峰期。针对这些问题,我们要对数据库系统进行整体优化,优化方面包括应用程序、数据库、主机和存储。为了更好地完成数据库系统的优化,我们将对系统进行多次的性能评估,收集相关的信息,然后对收集的信息进行评估,提出我们的解决方案。

数据库调优是一门艺术。调整完成之后,可以让数据库看起来更完美,用起来没有最快只有更快。

有些问题是一开始安装的时候,就没有安装好,参数也没有设置好。我们做调优的时候,尽可能地一并把这些问题解决掉。如果由于种种原因做不了的,也尽量让它的影响降到最低。最大限度地追求完美。做不了的,虽留有遗憾,也瑕不掩瑜。最后带给客户的是没有最快只有更快。

7.2 收集和了解哪些信息

数据库存在性能问题,我们就要寻找性能瓶颈,瓶颈一般又不是一个,需要从各个方面加以考虑。以下是我们需要了解的一些信息,有了这些信息,我们就可以知道调优从什么地方着手了。当然探寻这些问题的答案最好是调优工程师亲自登录系统去看，如果光远程传一个日志报告之类的，效率低不说，调优工程师也很难了解得很全面。除非是判断一个问题，只从 AWR 报告或执行计划中就可以大概地判定，整体地调优肯定是要现场操作。

（1）数据库的 AWR 报告、ASH 报告。
（2）操作系统的 CPU 利用率、内存使用率、I/O 响应时间、I/O 等待情况。
（3）数据库版本，PSU、CPU 补丁情况。
（4）数据库参数，包括隐含参数；操作系统参数。
（5）日志的组数，member 组员的大小，日志的写速度。
（6）高峰时间段的警告日志，相对应的 trace 文件。
（7）硬件资源是否充足，是否需要扩展，如增加内存。
（8）数据库 OLTP 的整体响应时间；报表批处理的响应时间。
（9）客户端期望的数据库响应时间；某个业务以前的响应时间。
（10）响应慢的应用主要等待时间耗在什么事件上。
（11）由于长年累积，数据段中的数据是否需要整理碎片，如表和索引的整理。
（12）内存充足的情况下，为了减少 I/O 等待，可以将表和常用的程序包 keep 到内存里。
（13）对于大表、大索引、绑定变量等问题，是否可以联合开发商进行操作。
（14）历史数据是否可以迁入历史库，减少核心库压力。
（15）Oracle 的自动任务，如审计、统计信息、直方图、auto space advisor、auto sql tuning 等。
（16）哪些 SQL 或程序运行较慢？SQL 是否有调优的空间？
（17）是否存在内存锁的争用？如 gc 开头的等待事件，latch free，library cache 等。
（18）热块问题。
（19）网络速度问题。

7.3 调优的依据和手段

为了把调优的手段介绍得更加全面，假设我们的数据库存在各种问题，都需要一一调整，目的是根据我们的判断，介绍更多的解决问题的方法。

1．redo 调整

增加 redo 日志组数，将 redo 由 RAID5 移至 RAID10，将 member 大小由 50MB 扩至 200MB。

判断的依据：

```
Top 5 Timed Events
~~~~~~~~~                                                    % Total
Event                                       Waits    Time (s) Ela Time
-----------------------------------------  --------  -------- ------
CPU time                                              94,674   39.25
db file sequential read                    6,098,736  65,327   27.08
log file sync                              1,293,986  43,891   18.20
```

```
db file scattered read                    2,092,560    24,769    10.27
buffer busy waits                           840,944     6,457     2.68
          -------------------------------------------

Wait Events for DB: YSDB   Instance: ysdb Snaps: 20064 - 20065
-> s  - second
-> cs - centisecond -   100th of a second
-> ms - millisecond - 1000th of a second
-> us - microsecond - 1000000th of a second
-> ordered by wait time desc, waits desc (idle events last)

                                                          Avg
                                              Total Wait  wait   Waits
Event                         Waits  Timeouts   Time (s)  (ms)   /txn
------------------------- ---------- -------- ---------- ------  -----
db file sequential read    6,098,736        0     65,327     11    4.0
log file sync              1,293,986    1,065     43,891     34    0.9
```

说明：

（1）log file sync 等待事件已经进入 top5，证明我们的 redo 写存在速度问题。

（2）log file sync 的等待时间，1 小时的快照间隔，它总共等待了 43 891 秒钟，12.2 小时，总的等待时间也是很长的，是一个可以提高速度的部分。

（3）再看平均等待时间 log file sync，平均等待时间为 34ms，而 Oracle 建议 log file sync 的平均等待时间不宜超过 5ms，由此大致可以推断，我们的 redo 日志是放在了 RAID5 上。建议调整到 RAID1 上。

（4）警告日志和 redo 相关的：

```
Fri May19 09:07:49 2015
Thread 1 cannot allocate new log, sequence 117398
Checkpoint not complete
... ...
Fri May 19 10:12:32 2015
Thread 1 cannot allocate new log, sequence 117421
Checkpoint not complete
```

警告日志报不能分配新的 redo 日志，检查点没有完成。也是由于磁盘的速度不好，redo 日志太小，日志组数不够三个方面可能都有责任。

（5）看一下 v$log：

```
SQL> select group#,thread#,bytes,members,status from v$log;

   GROUP#    THREAD#      BYTES  MEMBERS STATUS
---------- ---------- ---------- ------- ----------------
        1          1   52428800        2 ACTIVE
        2          1   52428800        2 ACTIVE
        3          1   52428800        2 CURRENT
```

说明：由于 I/O 速度慢，单位时间 redo 写的多，导致 redo 状态处于 active。如果底层做了 RAID，这里也没有必要使用两个 member 了。如果使用两个 member，必须把两个

member 分别放在不同的磁盘上。放在一个磁盘上,同时写,这不是给磁盘增加压力吗? 况且要坏都坏。如果非要防止人为删除,应该把两个 member 放在不同的挂载点下,不同的文件夹里。

以上的种种迹象表明,我们的 redo 需要调整。最优的方案是:❶将 redo 挪至 15000 转/分的磁盘做的 RAID1 上去。❷将日志组数增加至 6 组。❸将 redo 大小改为一个 400MB。然后再执行以上的检查。

调整的方法如下:

(1) 增加三个日志组至新的挂载点/ora_redo,这个挂载点是我们在存储上新划的 RAID1 磁盘。

```
SQL> alter database add logfile group 4 '/ora_redo/redo04.log' size 400m;
Database altered.
SQL> alter database add logfile group 5 '/ora_redo/redo05.log' size 400m;
Database altered.
SQL> alter database add logfile group 6 '/ora_redo/redo06.log' size 400m;
Database altered.
```

(2) 将当前的日志切换到 4,5,6 上的任意一个,然后稍等几分钟,等第 1、2、3 组日志不活动了,将 1、2、3 删除。

```
SQL> alter system switch logfile;
System altered.
SQL> /
System altered.
SQL> alter database drop logfile group 1;
Database altered.
SQL> alter database drop logfile group 2;
Database altered.
SQL> alter database drop logfile group 3;
Database altered.
```

(3) 1、2、3 组日志删除完成之后,再添加新的 1、2、3 组至/ora_redo。

```
alter database add logfile group 1 '/ora_redo/redo01.log' size 400m;
alter database add logfile group 2 '/ora_redo/redo02.log' size 400m;
alter database add logfile group 3 '/ora_redo/redo03.log' size 400m;
```

(4) 新的日志添加完成后,将旧的 1、2、3 组日志在物理上使用 rm 删除。

```
[oracle@yingshu ysdb]$ pwd
/u01/app/oracle/oradata/ysdb
[oracle@yingshu ysdb]$ rm -r redo*
```

至此我们的 redo 调整就完成了。我们重新查看新生成的 AWR 报告,再观察相应的等待事件,肯定有了很大的改观。警告日志也不再报无法分配新的 redo 了。

2. AIX5L、6.1 操作系统参数调整

运行在 IBM AIX 上的 Oracle 数据库调优过程中,相关的操作系统参数,主要有以下几

个需要关注。如果我们的生产库为 AIX 请关注以下几个参数。我这里直接给出了调整的命令,然后重新启动主机使参数生效。作用是让操作系统文件型内存使用的内存限制在 30% 以下,让更多的内存留给 Oracle 用做 ASM、裸设备、SGA,如果客户用的是文件系统,以下的值则不适用,需要另行评估计算参数的新值。

```
vmo – p – o maxclient% = 30
vmo – p – o maxperm% = 30
vmo – p – o minperm% = 5
vmo – p – o strict_maxperm = 1
```

3. Linux 操作系统参数调整

kernel.shmmax 参数:Linux 进程可以分配的单独共享内存段的最大值。一般情况下,该值应该大于 SGA_MAX_TARGET 或 MEMORY_MAX_TARGET 的值。如果主机上只运行一个实例,则可以设置为总内存大小的 65%。数据库占用的总内存大小也为 60% 左右。

kernel.shmmni 参数:设置系统级最大共享内存段的数量。Oracle 推荐值为 4096。

kernel.shmall 参数:设置共享内存的总页数,算法为总的物理内存大小除以一个分页的大小,这个值太小有可能导致数据库启动报错。分页大小由命令 getconf PAGE_SIZE 获取。对于主机内存为 96GB 的情况,该值的计算方法是:

```
[root@yingshu ~]# getconf PAGE_SIZE
4096
SQL> select 96 * 1024 * 1024 * 1024/4096 from dual;
96 * 1024 * 1024 * 1024/4096
---------------------------
                   25165824
```

semaphores 信号量是用来协调进程对共享资源的访问的,进程间访问共享内存时提供同步。

semmsl 参数:每个信号量组中信号量最大数量,推荐的最小值是 250。对于系统中存在大量并发连接的系统,推荐将这个值设置为 Processes 初始化参数加 10。

semmni 参数:系统中信号量组的最大数量。Oracle 10g 和 Oracle 11g 的推荐值为 142。

semmns 参数:系统中信号量的最大数量。semmns 的值不能超过 semmsl * semmni,超过了 semmsl * semmni 是非法的,因此推荐 semmns 的值就设置为 semmsl * semmni。oracle 推荐 semmns 的设置不小于 32000,假如数据库的 Processes 参数设置为 1500,则 semmns 的设置应为:

```
select (1500 + 10) * 142 from dual;
(1500 + 10) * 142
-----------------
           214420
```

semopm 参数:每次系统调用可以同时执行的最大信号量的数量。因为一个信号量组最大拥有 semmsl 个,因此可以将 semopm 设置为 semmsl 的值。Oracle 验证的 Oracle 10g

和 Oracle 11g 的 semopm 的配置为 100。

修改以上参数的方法，可以用 vi 编辑器编辑/etc/sysctl.conf 这个文件，然后使用 sysctl -p 命令使其生效。

4. 数据库参数调整

从 Oracle 10g 开始，内存可以自动管理（Automatic Shared Memory Management，ASMM）了，我们只需要设一个 sga_target、sga_max_size 或 memory_target、memory_max_target 就不用管下面各个池的大小了。但是，自动管理内存好吗？事实证明，内存自动管理在数据库压力比较大的情况下，会使分配给某个池的内存一会大一会小，容易出事。所以我不建议大家用内存自动管理，还是手工管理内存更加稳定。

Oracle 10g 内存自动管理还受 statistics_level 参数的影响，该参数必须设置为 typical 或 all，内存自动管理才生效，如果此参数的值是 basic，则内存自动管理无效。

Oracle 10g 内存自动管理可以管理 5 个内存池的大小，它们是 buffer cache、shared pool、large pool、java pool 和 stream pool。其他的内存参数仍然需要手工设置。视图 v$sga_dynamic_components 记录了可以动态调整的内存的大小。通过这个视图可以查询当前的自动设置的值。

内存自动管理受后台进程 MMAN（Memory Manager）的控制。由此后台进程进行内存建议的收集，收集来建议之后，也是由此进程自动调置对应的参数。它设置的参数都是以两个下划线开始，如"__shared_pool_size"是自动调整的参数。数据库启动时，会比较 DBA 设置的参数和自动调整的参数，shared_pool_size 设置的比"__shared_pool_size"大，则用 shared_pool_size，否则用"__shared_pool_size"。

我的习惯是不用内存自动管理，建库的时候就不用内存自动管理键，使用手工设置内存的大小，省得内存自动管理在以后的运行中出故障。如果目前数据库内存为自动管理，我们也建议改为手工管理，下面介绍一下内存自动管理改为手工管理的方法。

为了避免修改的过程中出现错误致数据库起不来，我们先备份一下当前的 spfile。拷贝也行，创建一个 pfile 到某个目录下也行。

```
SQL> create pfile = '/tmp/pfile20150512' from spfile;
File created.
```

主机的总内存是 160GB，我们设置的各个内存池的大小，都跟总内存有关系，可以根据自身的情况再做一下大小的微调，总的比例应该是在 sga 中 db_cache_size 用于缓存数据的内存区域最大，用于存放数据字典和 SQL 语句的 shared_pool_size 内存区域第二大。如果此主机上只运行一个实例，那么 sga 占主机总内存的 60% 左右。

```
alter system set sga_max_size = 120g scope = spfile sid = '*';
alter system set sga_target = 0 scope = spfile sid = '*';
alter system set db_cache_size = 80g scope = spfile sid = '*';
alter system set shared_pool_size = 10g scope = spfile sid = '*';
alter system set java_pool_size = 120m scope = spfile sid = '*';
alter system set large_pool_size = 500m scope = spfile sid = '*';
alter system set log_buffer = 79193088 scope = spfile sid = '*';
alter system set pga_aggregate_target = 32g scope = spfile sid = '*';
```

5. 将常用的应用程序包 keep 到内存中

对于频繁使用的数据库对象而言，如存储过程、触发器、序列、游标等，将其 keep 到内存中，可以减少物理 I/O 提高用户的响应时间。keep 的过程需要一个软件包 dbms_shared_pool，这个软件包在 Oracle 11g 中不是被默认创建的，需要执行 dbmspool.sql 来创建。有了 dbms_shared_pool 这个包之后，用户就可以将自己常用的程序 keep 到内存中了。

我们可以先从 v$db_object_cache、v$sqlarea 里将执行次数多的程序包查询出来。然后根据结果判断，哪些包是我们最常用的，是需要 keep 的。被经常使用的程序名用下面的语句查询，其中执行的次数，以及包占用内存的大小可以做适当的调整。

```
select a.OWNER,                         //对象的 owner
    a.name,                             //对象的名称
    a.sharable_mem,                     //共享内存占用大小
    a.kept,                             //是否被 keep 在内存中
    a.EXECUTIONS ,                      //执行的次数
    b.address,                          //语句的地址
    b.hash_value                        //hash 值
from v$db_object_cache a,
    v$sqlarea b
where a.kept = 'NO' and
(( a.EXECUTIONS > 1000                  //执行次数
and a.SHARABLE_MEM > 50000)             //共享内存占用大小
or a.EXECUTIONS > 10000)                //执行次数
and SUBSTR(b.sql_text,1,50) = SUBSTR(a.name,1,50);
```

如果了解我们的应用程序，知道哪个用户下的哪些包会被经常使用，也可以通过用户名、程序名确定我们要 keep 的内容。用的视图就是 dba_objects。

```
select object_name,object_type from dba_objects where owner = 'YS';
```

keep 对象之前我们先使用 dbmspool.sql 创建一下 dbms_shared_pool 包。使用 sys 用户以 DBA 身份执行。

```
SQL> @?/rdbms/admin/dbmspool.sql
Package created.
Grant succeeded.
```

创建完成了之后，我们就可以把常用的程序包放在内存里了。例如通过上面的查询判断，我们得出以下的操作系统包被经常使用，这些包属于 sys 用户，用得很频繁，就可以用下面的命令将它们 keep 到内存中。如果是应用用户的程序，keep 的方法也一样。

```
exec dbms_shared_pool.keep('STANDARD');
exec dbms_shared_pool.keep('DBMS_SYS_SQL');
exec dbms_shared_pool.keep('DBMS_SQL');
exec dbms_shared_pool.keep('DBMS_UTILITY');
exec dbms_shared_pool.keep('DBMS_DESCRIBE');
exec dbms_shared_pool.keep('DBMS_JOB');
exec dbms_shared_pool.keep('DBMS_STANDARD');
exec dbms_shared_pool.keep('DBMS_OUTPUT');
```

```
exec dbms_shared_pool.keep('DIANA');
exec dbms_shared_pool.keep('PIDL');
```

6. 将常用的小表 Keep 到内存中

我们的应用开发商应该知道哪些表比较常用；通过 AWR 报告里的物理读(Segments by Physical Reads)也可以得知哪些段比较消耗物理 I/O；我们可以通过 dba_segments 查询到这个表有多大。如果大小合适，内存足够放得下，这时，如果 I/O 又成为我们性能的瓶颈，就可以把对应的表 keep 到内存中来。一旦实现 keep，省去了 I/O 的时间，我们的 SQL 执行速度将明显提升。就算存储的性能不好，也没有关系，因为除第一次读取，我们这些表从此不在存储里读取了。

```
......
Segments by Physical Reads       DB/Inst: YSDB/ysdb     Snaps: 10762-10763
-> Total Physical Reads:          1,572,493
-> Captured Segments account for   38.9% of Total

                Tablespace                     Subobject  Obj.    Physical
Owner           Name        Object Name        Name       Type    Reads   % Total
--------------- ----------- ------------------ ---------- ------- ------- -------
YS              YS_DATA01   YS_KUCUNZ                     TABLE       114   35.74
YS              YS_INDEX01  IDX_CAIWUZ                    INDEX         6    1.88
YS              YS_INDEX01  PK_YS_RENINFO                 INDEX         2     .63
YS              YS_INDEX01  IDX_RENINFO_PWD               INDEX         1     .31
SYS             SYSAUX      WRI$_ADV_OBJECTS              TABLE         1     .31
                            -------------------------------------------
......
```

上面的 AWR 报告显示，表 YS.YS_KUCUNZ 占用的 I/O 最大，物理读最频繁。我们查看一下它有多大：

```
SQL> col segment_name for a15
SQL> col segment_type for a15
SQL> select owner,segment_name,segment_type,bytes/1024/1024 from dba_segments where segment_name = 'YS_KUCUNZ';
OWNER           SEGMENT_NAME    SEGMENT_TYPE    BYTES/1024/1024
--------------- --------------- --------------- ---------------
YS              YS_KUCUNZ       TABLE                   80.0625
```

ys.ys_kucunz 这张表有 80MB，不算太大，可以将它 keep 到内存中。keep 表的语句如下：

```
SQL> alter table ys.kucunz storage(buffer_pool keep);
Table altered.
```

如果还有其他的表需要常驻内存，方法和上述的一样。

7. 大表重组

日积月累的表的碎片，也是我们要调整的对象。经过几年的增删改操作，表里已经积累了大量的碎片。查看表的碎片之前，我们先收集一下表的统计信息，统计信息不准确，我们查的碎片程度也不准确。

```
SQL> exec dbms_stats.gather_table_stats(ownname =>'SHU',tabname => 'STUD');
PL/SQL procedure successfully completed.
```

可以用以下的语句来查询表的碎片程度。浪费空间比较多的表,就是我们要重组的重点了。

```
select table_name,                                          //表名
round((blocks * 8), 4) "highwm_k",                          //高水位线占用的大小
round((num_rows * avg_row_len / 1024), 4) "real_use_k",     //真实的大小
round((blocks * 10 / 100) * 8, 4) "free_spc_(pctfree) k",   //有多少预留空间
round((blocks * 8 - (num_rows * avg_row_len / 1024) - blocks * 8 * 10 / 100), 4) "waste_k"
from dba_tables                                             //(上一行)浪费的空间
where table_name = 'STUD';                                  //想了解碎片的表名
TABLE    HighWM_k  Real_Use_k  Free_Spc_(Pctfree) k    Waste_k
-----    --------  ----------  --------------------    -------
STUD     11024     0           1102.4                  9921.6
```

对于数据变化量比较大的表,我们可以对其进行重组。重组的方法有多种。

(1) expdp 导出来,将原表删除,impdp 再导进去,实现了对表碎片的整理。这种办法做数据备份迁移的时候常用,整理碎片一般不用这个方法。

(2) create table TEACH2 as select * from teach;这种方法也不怎么常用,因为它生成一个新表,还要修改表名,又不能将表中的默认值插到新表里去,用的时候需要谨慎。

(3) 第三种方法是 move 表,就是我们最常用的方式,执行的时间一定是在业务不忙的时候,保证表空间的剩余空间要比此表大。如果表空间没有剩余空间可以将表 move 到其他表空间,再 move 回来。对于特别大的表或分区,一定要计算出大概的 move 时间,在规定的停业务时间必须完成操作,不能因为 move 时间过长,影响了营业。注意:move 表之后,此表上面所有的索引都将失效,需要重建所有此表上的索引。

```
SQL> alter table ys.teach move;
Table altered.
```

将表 move 至其他表空间,再 move 回来:

```
SQL> alter table ys.teach move tablespace users;
Table altered.
SQL> alter table ys.teach move tablespace ystbs;
Table altered.
```

如果是 move 分区表,需要一个分区一个分区地做:

```
SQL> alter table ys.stud move partition part_001;
Table altered.
SQL> alter table ys.stud move partition part_002;
Table altered.
```

(4) shrink table

从 Oracle 10 开始,又推出了一种新的整理表碎片的方法,就是 shrink 收缩。这个功能

必须打开行移动,有三个不同的选项表示三种不同的整理级别。shrink 不需要重建索引,是比 move 进步的一点,但 shrink 会改变数据的 rowid,如果应用在使用的过程中调用了 rowid,那么就不能用 shrink 了,使用 LOGMINER 恢复过程中也会用到 rowid,看来 shrink 还会影响 LOGMINER,但以上两种假设在我们的实际生产中很少用到,shrink 还是可以用的。

打开行移动:

```
SQL> show user
USER is "SHU"
SQL> alter table stud enable row movement;
Table altered.
```

收缩表之前先收集一下表的统计信息。统计信息里有表的高水位线,我们在查询高水位线之前必须先收集统计信息,否则,我们的高水位线有可能是不准确的:

```
SQL> exec dbms_stats.gather_table_stats(ownname =>'SHU',tabname => 'STUD');
PL/SQL procedure successfully completed.
```

我们先了解一下 High Water Mark 的计算方法,高水位线的概念。

查询表的高水位线:

```
SQL> select blocks, empty_blocks,num_rows from dba_tables where table_name = 'STUD' and owner =
'SHU';
    BLOCKS EMPTY_BLOCKS   NUM_ROWS
---------- ------------ ----------
       370            0     100000
```

结果显示,表里没有空块,这个表不需要我们整理。

为了达成我们的实验环境,我们将表里的数据删除一半:

```
SQL> delete from stud where studid <= 50000;
50000 rows deleted.
```

查询在高水位线以下,有多少 MB 空闲空间:

```
select table_name,
(blocks * 8192 / 1024 / 1024) -
(num_rows * avg_row_len/ 1024 / 1024) "data lower than hwm in mb"
from user_tables
where table_name = 'STUD';
TABLE_NAME              Data lower than HWM in MB
------------------      -------------------------
STUD                    2.17536926
```

```
SQL> alter table stud shrink space compact;
Table altered.
```

收缩表,但高水位线保留。

```
SQL> select blocks, empty_blocks,num_rows from dba_tables where table_name = 'STUD' and owner =
'SHU';
Used Blocks EMPTY_BLOCKS NUM_ROWS
---------- ------------ --------
       370            0    50000
```

高水位线并没有降低。

```
SQL> alter table stud shrink space;
Table altered.
```

收缩表,降低高水位线。

```
SQL> alter table stud shrink space cascade;
Table altered.
```

收缩表,降低高水位线,相关索引也收缩一下。
再收集一下表的统计信息。

```
SQL> exec dbms_stats.gather_table_stats(ownname =>'SHU',tabname => 'STUD');
PL/SQL procedure successfully completed.
```

再看表的高水位线,已经降下来了。

```
SQL> select (b.blocks - a.empty_blocks - 1) hwm
from user_tables a,
     user_segments b
where a.table_name = b.segment_name and
    a.table_name = 'STUD';
    HWM
----------
     46
select segment_name,blocks from user_segments where segment_name = 'STUD';
```

8. 整理索引

索引碎片多了之后,也会发生性能问题,我们在 AWR 报告里看到 db file sequential read 等待事件,通常和效率不高的索引有关。如果我们的索引被删除的数据超过 20%,或 BLEVEL 大于等于 4,则我们的索引就需要整理了。补充一句,任何一种方法整理索引,都不应该在业务生产时间,都应该在停业时间,系统不繁忙的时间进行。也就是你要确认这个时间段此索引无人使用才可以整理。

查询数据库中 blevel 大于等于 4 的索引:

```
select owner || '.' || index_name as "owner.index_name",blevel
from dba_indexes
where blevel >= 4
order by blevel desc;
```

如果被删除的行 del_lf_rows/lf_rows * 100 大于 20%,则此索引建议整理:

```
col name for a20
select name,blocks,del_lf_rows/lf_rows * 100,del_lf_rows/lf_rows * blocks from index_stats;
```

如果索引的状态为 UNUSABLE,则需要整理索引:

```
select owner,index_name,status from dba_indexeswhere status = 'UNUSABLE';
```

在整理索引之前,我们需要看一下索引的 owner 是谁,只有和我们应用相关的才需要整理。

整理索引的方法有以下几种。

(1) 先准备好建立索引的脚本,然后将索引删除,重建索引。

```
select index_name,dbms_metadata.get_ddl('INDEX',INDEX_NAME,TABLE_OWNER) index_ddl
from user_indexes
where table_name = 'STUD';

drop index shu.ind_stud_id;

create unique index shu.ind_stud_id on shu.stud ("STUD_ID");
```

这种方法速度快,好理解。影响关于此索引的 dml。在实际生产中,可以使用。

(2) rebuild,扫描现有索引块进行索引重建,需要两倍索引大小的空间,降低高水位线,执行速度快,会全程对表加锁,阻塞 DML 操作。是最常用的整理索引的方法。

```
alter index shu.ind_stud_id rebuild;
```

(3) rebuild online 对表进行全扫实现索引重建,会降低索引高水位线,一般执行速度慢,只在开始和结束时对表加锁,执行中间不阻塞 DML 操作。因为速度比较慢,中间过程又要维护 DML 的日志表,如果有 DML 操作,那么整理时间将大大拉长。所以谨慎使用。

```
alter index shu.ind_stud_id rebuild online;
```

(4) coalesce 接合索引,只对索引块做合并操作,不下降高水位线,可随时中断。全程不阻塞 DML 操作。因为会产生更多的 redo 日志。

```
alter index shu.ind_stud_id coalesce;
```

(5) shrink 索引,原理请参见 shrink table。

```
alter index shu.ind_stud_id shrink space;
alter index shu.ind_stud_id shrink space compact;
alter index shu.ind_stud_id shrink space cascade;
```

7.4 性能优化的定义和范围

本次性能优化的服务范围为樱澍保险的财险核心生产系统。

工程师将从应用层、数据库层、操作系统层到存储层对樱澍的核心生产系统进行全面的优化咨询服务。

本次数据库的优化可能涉及对应用 SQL 进行调整,并针对应用中存在的待改进的问题提出解决建议。在优化过程中部分对应用的调整需要开发商的配合。

7.5 性能优化的目标

数据库的响应速度主要体现在数据库的等待时间和响应时间上。体现在如下几个方面:

- 响应时间(Response Time);
- CPU 时间(Service Time);
- 等待时间(Wait Time);
- 查询量(User Calls);
- 事务量(transactions);
- Redo Size/秒;
- 系统 CPU(usr+sys%);
- 系统 IO 等待(wio%);
- 系统内存使用;
- 系统总体性能提升 25% 以上;
- 访问最少的块数;
- 将数据块保留在内存中。

7.6 深入研究数据库系统的五大资源

这五大资源和性能息息相关,需要一一排查是否存在性能瓶颈。
(1) 内存;
(2) 程序(SQL);
(3) I/O;
(4) CPU;
(5) 网络。

7.7 性能优化需要考虑的问题

(1) 测量并记录当前性能。
(2) 确定当前 Oracle 的性能瓶颈(等待什么)。
使用 awrrpt,sqlrpt,addmrpt,sqltrace,ashrpt,tuning 包等。
操作系统工具如 top,sar,vmstat,iostat,osw 等。
(3) 确定当前的 OS 是否有瓶颈,若有则分析瓶颈的原因。
(4) 操作系统核心参数的调整建议,包括信号量、内存分配限制。
(5) 数据库内存配置的调整建议。

包括 share pool，db_cache_size，keep 池，recycle 池，写进程数等。尽量使用手动管理内存大小，避免内存忽大忽小，减少抖动。

（6）I/O 配置的调整。分散 I/O，redo 使用 RAID10，热表 cache 到内存里，数据文件响应时间。

（7）清理不需要的历史数据，或把历史数据导入历史库。

（8）分析主要业务表，为优化器的工作提供足够准确的统计分析数据，使 CBO 优化器能够生成优化的执行计划。关闭自动统计信息。

（9）使用 keep 池，将常用的表或其他对象 keep 到内存里来。

（10）使用 rebuild online 重建重点业务表索引。

（11）keep 常用应用程序包。

（12）Keep 常用小表。

（13）大表重组。

（14）关键索引重建。

（15）制订历史数据处理细则。例如将一年前的数据导出生产库，导入历史库。

（16）将特大的生库表且又不能导入历史库的大表考虑进行分区。例如按月分区，或按（半）年分区；按列表分区。

（17）确保代码可共享。为了优化共享的 SQL，使用绑定变量而不是文本值。

（18）尽可能避免或者减少排序操作。尽可能确保排序在内存中执行，减少 temp 表空间使用量。

（19）减少分页和交换。

（20）应用分别分布于 RAC 两个或多个节点，避免交叉访问。可以考虑禁用 DRM。

（21）每一阶段调优完成，记录当前性能，与初期的(上一阶段的)性能做前后比较，量化调优收益。

7.8 风险防范措施

（1）确保数据库备份完整可用。

（2）所有操作和检查环节都使用事前完成并预演测试通过的脚本，避免临时修改脚本。

（3）每部分完成，通过检查确认无误，再进行其他部分，避免互相干扰。

（4）Oracle 工程师和应用系统专家现场支持，及时处理突发问题。

7.9 数据库优化结果的保持

（1）在数据库系统没有大的变动的情况下，在服务期内，甚至更长的一段时间，通过定期的维护及相应的维护手段保证数据库优化后性能稳定。

（2）定期清理表中的垃圾数据或导入历史库，协助制定常用表的相关操作维护计划。

（3）优化表结构。

（4）减少操作语句间的相互影响(如锁表问题)。

（5）提高表操作语句的执行效率。

7.10 客户收益

(1) 提高投资回报率

合理、充分利用资源,延长系统使用寿命;

优化运维成本,降低 IT 系统管理成本。

(2) 提高系统稳定性

在问题发生之前就发现并解决问题,因而可做到高枕无忧。

(3) 提高运维水平

系统优化的知识与经验能够得到传递,提高运维人员的能力。

(4) 提高企业核心竞争力

市场需求适应性增加,缩短新业务上线周期,提高 IT 系统的灵敏度,适应了飞速发展的业务需求。

(5) 系统运行速度更快,可用性更高,客户满意度升高,企业竞争力提升。

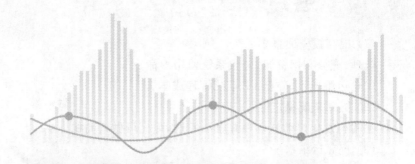

第 8 章 Oracle数据库的迁移

8.1 概述

在日常工作中,常常会遇到数据迁移的需求。例如:搭建测试环境,更换新的硬件设备,版本的升级,在不同类别的数据库间迁移等情况。需求不同,制定的方案也不同,但迁移的方法就这么几种,再多种的需要,最后无外乎使用这些方法。

8.2 常用的数据库迁移方法

- 数据量比较小的情况下,可以直接利用数据库工具 exp、imp。
- 数据量较大时,可以利用 Oracle exp、imp 结合操作系统 pipe 管道技术并发多个进程进行导出导入。
- 数据库大时,可以利用建立数据库链 link。按表在旧库 select,在新库 insert append 的方式。此方式比 exp、imp 速度快。
- 使用 RMAN 在源数据库上备份,在目标数据库上实现恢复,实现迁移的目的。
- 将 spfile、数据文件、控制文件、redo 日志复制至新配置的环境,实现迁移。
- 利用 ASM 加盘减盘技术实现数据库迁移,对于更换存储是特别好的方案。
- 利用第三方厂商技术进行数据复制,如 DSG、SharePlex。
- 使用存储级的复制裸设备或数据文件实现迁移。
- 使用 DataGuard 或 GoldenGate 实现数据无缝割接迁移。
- 使用 SQL loader 工具。
- 使用外部表,将少量数据加载到数据库中。
- 使用 convert database to platform 'Linux 64-bit for AMD' 实现不同平台的迁移。

8.3 迁移方案一

8.3.1 概述

为适应目前日益增长的数据量,以及确保数据库的稳定正常运行,应北京樱澍科技有限公司的邀请,我方充分调研了运行现状,了解了具体状况,并给出关于 Oracle 数据库的迁移和升级方案。

现计划将①樱澍集团管控平台、②集团合并报表、③集团协同办公平台这 3 个数据库迁移整合到一个 RAC 数据库。其中①、③应用在节点一运行,②应用在节点二运行。

8.3.2 编写目的

解决目前数据库运行不稳定的问题,实现集中管理,节约硬件成本,减少运维支出,目前有数据库运行在 Windows 上。Windows 操作系统本身的稳定性远远低于 AIX,给数据库造成潜在宕机或丢失数据库的危险。迁移到新环境升级改造完成后会明显降低数据量对各个环节的压力,给应用提供稳定的后台保障。

8.3.3 迁移时间

确定停机时间,安排 6h 停机时间,19:00—01:00。因升级需停监听器,关闭应用服务。迁移时间安排在事务量较小的周六进行。

8.3.4 数据库迁移规划方案

根据对此次迁移环境的了解,决定使用逻辑的导入导出(expdp/impdp)完成此次数据库的迁移。

这种方式适用于小数据量的数据迁移,使用数据泵直接导出数据结构和数据,速度较快,简单方便。而且逻辑的导入导出可以跨操作系统平台和数据库版本,本次的迁移牵扯到操作系统平台和数据库版本,这种方式恰好可以满足需求,但是这种方式需要一定的停机时间,按照现有生产数据库的数据量大概需要 4h 的时间(时间还受网络带宽和服务器性能的影响),安排 6h 停机,把工作量不能安排得太紧。

8.3.5 迁移前的准备

1. 搭建新环境

(1) 稳定优化的新环境是数据库迁移后正常运行的重要保证。

(2) 系统规划是长期稳定运行的一个重要条件,实施前一定要规划好,不能在无方案的情况下实施,边干边改。具体要注意的事项参见第 3 章。

(3) 本次的新环境为 AIX 操作系统,AIX 6.1 的操作系统补丁一定要打全,最好补丁升级到最新的补丁,以保障操作系统的稳定性,也减少了再次升级补丁给数据库带来隐患。同时也建议参考操作系统专家的建议。

2. 搭建过程

(1) 连接硬件,安装操作系统,配置 IP,配置主机名等。

(2) 安装 Oracle 数据库 11.2.0.4.7 for AIX。详见第 4 章。

(3) 在 support.oracle.com 上找到相应的版本升级补丁并打齐。

(4) DBCA 创建数据库,创建相关表空间,创建 index 表空间,创建所需要的业务用户,例如以下命令:

```
create tablespace data
datafile '+DGDATA'
size 6000m
autoextend off;

create tablespace index
datafile '+DGDATA'
size 6000m
autoextend off;

create user orcl identified by oracle
default tablespace data
temporary tablespace temp
quota unlimited on index;
```

3. 关闭新建数据库的归档模式

```
alter database noarchivelog;
```

4. 创建 expdp/impdp 目录

源端数据库实例: `create directory exp_dp as '/backup/expdp';`
新数据库实例: `create directory exp_dp as '/backup/impdp';`

5. 调整源端数据库系统的 JOB

调整源端数据库 JOB 自动运行时间,以免在导出数据过程中 JOB 自动运行。

8.3.6 迁移过程

1. 停止源端数据库服务

周六 19 点整,停止源端数据库应用和监听;重启数据库实例;检查 objects:

```
lsnrctl stop
srvctl stop database -d ysdb
srvctl start database -d ysdb
```

检查有没有业务的定时任务:

```
select * from dba_jobs;
```

检查当前应用用户 user_objects,以便迁移完成检查对象个数:

```
select count(*) from user_objects;
select count(*) from user_objects where object_type = 'TABLE';
select count(*) from user_objects where object_type = 'INDEX';
```

```
select count( * ) from user_objects where object_type = 'VIEW';
select count( * ) from user_objects where object_type = 'MATERIALIZED VIEW';
select count( * ) from user_objects where object_type = 'TRIGGER';
select count( * ) from user_objects where object_type = 'SEQUENCE';
select count( * ) from user_objects where object_type = 'PROCEDURE';
select count( * ) from user_objects where object_type = 'DATABASE LINK';
select count( * ) from user_objects where object_type = 'LOB';
```

2. 源端数据库导出

```
nohup expdp system/****** dumpfile = expdp.dmp schemas = ****,****,***** logfile = expd.log directory = exp_dp parallel = 4&
```

导出成功后,关闭数据库:

```
srvctl stop database -d db
```

检查 impdp.log 日志中是否有警告和错误:

3. 将导出的 dump 文件传到新环境/backup/impdp

```
sftp > get expdp.dmp
```

4. 在新环境中导入数据

```
nohup impdp system/**** directory = exp_dp dumpfile = expdp.dmp logfile = impdp.log schemas = ****,****,*****&
```

检查 impdp.log 日志中是否有警告和错误。

5. 检查 JOB 是否成功导入

```
select * from dba_jobs;
```

6. 检查 user_objects 个数

```
select count( * ) from user_objects;
select count( * ) from user_objects where object_type = 'TABLE';
select count( * ) from user_objects where object_type = 'INDEX';
select count( * ) from user_objects where object_type = 'VIEW';
select count( * ) from user_objects where object_type = 'MATERIALIZED VIEW';
select count( * ) from user_objects where object_type = 'TRIGGER';
select count( * ) from user_objects where object_type = 'SEQUENCE';
select count( * ) from user_objects where object_type = 'PROCEDURE';
select count( * ) from user_objects where object_type = 'DATABASE LINK';
select count( * ) from user_objects where object_type = 'LOB';
```

7. 打开数据库归档

```
alter database archivelog;
```

8. 关注迁移后的性能状况和表的统计信息

第 9 章 OCM考试练习实验

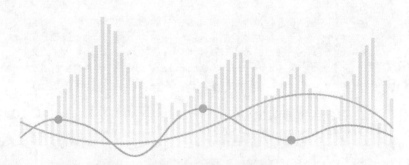

OCM(Oracle Certified Master)是 Oracle 公司的最高认证考试。曾经受到不少技术爱好者的追捧,进入中国市场之后,经过几年的发展,已经量产。建议把以下的实验做熟练。考不考 OCM 认证,自己看着办吧,技术好的话,可以不用考。OCA、OCP 是背题库,考试的时候都是题库中的原题,背背题库就能考 90 分以上,和技术好坏没关系。OCM 是比着题库做实验,比 OCP 有难度。

9.1 手工建库

使用 CREATE DATABASE 命令,按下面的要求创建一个数据库。

❶ 数据库(Database)的名字是 PROD。

❷ 实例(Instance)的名字是 PROD。

❸ 设置 ORACLE_SID 环境变量。//这里很奇怪,使用.bash_profile 设置不生效,必须使用 export ORACLE_SID=PROD。

❹ SYS 和 SYSTEM 用户的密码都设置为 ORACLE。

❺ 在/home/oracle/scripts 目录下,有一个参数文件名为 initPROD.ora。请按需求修改它,便于手工建库时使用。

❻ 创建一个本地管理的(Locally Managed) SYSTEM 表空间。

❼ 在目录/u01/app/oracle/oradata/PROD 下,使用目录结构(disk1-disk5),模拟在不同的磁盘上分散物理文件。

❽ 不要执行其他的数据库脚本文件,如 catalog.sql,catproc.sql 等。

注：不要创建其他的数据库,保证数据库处于打开状态。

```
su - oracle
export ORACLE_SID = PROD                              //输出环境变量
-- .bash_profile
alias ss = "sqlplus '/as sysdba'"                     //设置 SQLPLUS 的别名
```

```
-- mkdir                              //建对应的目录
mkdir -p /u01/app/oracle/admin/PROD/adump
mkdir -p /u01/app/oracle/admin/PROD/bdump
mkdir -p /u01/app/oracle/admin/PROD/cdump
mkdir -p /u01/app/oracle/admin/PROD/udump
mkdir -p /u01/app/oracle/oradata/PROD/disk1
mkdir -p /u01/app/oracle/oradata/PROD/disk2
mkdir -p /u01/app/oracle/oradata/PROD/disk3
mkdir -p /u01/app/oracle/oradata/PROD/disk4
mkdir -p /u01/app/oracle/oradata/PROD/disk5

-- initPROD.ora                       //整理参数文件
cat init.ora | grep -v ^# | grep -v ^$ > initPROD.ora
add as flow
db_name = PROD
sga_max_size = 280m
sga_target = 280m
control_files = ("/u01/app/oracle/oradata/PROD/disk1/controlfile01.dbf",
                 "/u01/app/oracle/oradata/PROD/disk2/controlfile02.dbf",
                 "/u01/app/oracle/oradata/PROD/disk3/controlfile03.dbf")
*.compatible = '10.2.0.1.0'
*.db_block_size = 8192
*.db_domain = ''
*.db_file_multiblock_read_count = 16
*.log_archive_dest_1 = 'location=/u01/app/oracle/oradata/arch'
*.log_archive_format = '%t_%s_%r.dbf'
*.open_cursors = 300
*.pga_aggregate_target = 83886080
*.processes = 500
*.remote_login_passwordfile = 'EXCLUSIVE'

-- startup nomount                    //启动到 nomount 状态
startup nomount pfile='$ORACLE_HOME/dbs/initPROD.ora'

-- create spfile
create spfile from pfile;
startup nomount force;

-- modify                             //修改参数
alter system set background_dump_dest = '/u01/app/oracle/admin/PROD/bdump' scope = spfile;
alter system set core_dump_dest = '/u01/app/oracle/admin/PROD/cdump' scope = spfile;
alter system set user_dump_dest = '/u01/app/oracle/admin/PROD/udump' scope = spfile;
alter system set audit_file_dest = '/u01/app/oracle/admin/PROD/adump' scope = spfile;
alter system set db_create_file_dest = '/u01/app/oracle/oradata/PROD/disk3' scope = spfile;
alter system set db_create_online_log_dest_1 = '/u01/app/oracle/oradata/PROD/disk1' scope = spfile;
alter system set undo_management = auto scope = spfile;
alter system set undo_tablespace = undotbs scope = spfile;
alter system set job_queue_processes = 5 scope = spfile;
```

```
startup nomount force;

-- password file                                          //建密码文件
orapwd file = orapwPROD password = oracle entries = 5 force = y

-- create database                                        //手工建库
create database prod
    user sys identified by oracle
    user system identified by oracle
    logfile group 1 ('/u01/app/oracle/oradata/PROD/disk1/redo01.log') size 100m,
            group 2 ('/u01/app/oracle/oradata/PROD/disk1/redo02.log') size 100m,
            group 3 ('/u01/app/oracle/oradata/PROD/disk1/redo03.log') size 100m
    maxlogfiles 30
    maxlogmembers 5
    maxloghistory 1
    maxdatafiles 500
    maxinstances 2
    character set utf8
    national character set utf8
    datafile ' /u01/app/oracle/oradata/PROD/disk2/system01. dbf ' size 325m reuse extent
management local
sysaux datafile '/u01/app/oracle/oradata/PROD/disk3/sysaux01.dbf' size 325m reuse
    default temporary tablespace tempts1
        tempfile '/u01/app/oracle/oradata/PROD/disk4/temp01.dbf' size 20m reuse
    undo tablespace undotbs
        datafile '/u01/app/oracle/oradata/PROD/disk5/undotbs01.dbf'
        size 200m reuse autoextend on maxsize unlimited;

@?/rdbms/admin/catalog.sql                                //创建数据字典
@?/rdbms/admin/catproc.sql                                //创建基本的过程和包
```

9.2 数据库设置和 undo 管理

(1) 在数据库 PROD 中，执行最少需求的脚本，完成最基本的配置需求。

(2) 设置自动 undo 管理，支持下面的需求：

❶ 避免 ORA-01555 快照太旧错误，支持平均查询时间 90 分钟。

❷ 通常的业务时间段，并行的 OLTP 用户大约为 120 个。

❸ 在晚上和周末并行的批处理进程达到 15 个。

```
@?/rdbms/admin/catblock.sql                               //创建有关 lock 的视图
@?/rdbms/admin/catoctk.sql                                //关于安全的数据字典
@?/rdbms/admin/owminst.plb                                //定制数据库相关
conn system/oracle
@?/sqlplus/admin/pupbld.sql                               //创建 product_user_profile table
@?/sqlplus/admin/help/hlpbld.sql helpus.sql               //安装 SQLPLUS 帮助
```

```
alter system set undo_retention = 5400 scope = both;        //undo 保留时间

-- undo guarantee
alter tablespace undotbs retention guarantee;                //保护 undo 信息不被轻易读写
alter tablespace undotbs retention noguarantee;
alter system set processes = 200 scope = spfile;             //修改 processes 参数
alter system set transactions_per_rollback_segment = 15 scope = spfile;
```

9.3 创建 listener

使用默认的 listener 名创建一个 listener。
(1) 使用 TCP/IP 协议,使用主机名(不使用 IP 地址)。
(2) 这个 listener 监听默认的端口号。
(3) 数据库 PROD 和 EMREP 将由这个 listener 提供服务。
增加第 2 个 listener,名为 LSNR2,使用 1526 端口,支持实例注册。
❶ 设置 PROD 实例自动注册到 LSNR2 监听程序中。
❷ 同时启动这两个 listener。

```
-- 1.1 listener.ora
在 $ ORACLE_HOME/network/admin 下. 手工编辑 listener,静态监听 prod 和 emrep 实例.
-- listener.ora
SID_LIST_LISTENER =
  (SID_LIST =
    (SID_DESC =
      (SID_NAME = PROD)
      (ORACLE_HOME = /u01/app/oracle/product/10.2.0/db_1)
    )
    (SID_DESC =
      (SID_NAME = EMREP)
      (ORACLE_HOME = /u01/app/oracle/product/10.2.0/db_1)
    )
  )
创建两个 listener,另一个 listener 名叫 lsnr2.
LISTENER =
  (DESCRIPTION_LIST =
    (DESCRIPTION =
      (ADDRESS = (PROTOCOL = TCP)(HOST = yingshu)(PORT = 1521))
    )
  )

LSNR2 =
  (DESCRIPTION_LIST =
    (DESCRIPTION =
      (ADDRESS = (PROTOCOL = TCP)(HOST = yingshu)(PORT = 1526))
    )
```

```
)
```
将数据库的默认 listener 设置成 lsnr2
```
alter system set local_listener = '(address = (protocol = tcp)(host = yingshu)(port = 1526))'
scope = both;

alter system set local_listener = 'LSNR2' scope = spfile;
alter system register;
```

9.4 共享服务配置

(1) 配置 PROD 数据库支持 300 个会话,100 个专有服务(Dedicated)连接。
(2) 配置 PROD 数据库支持如下要求:
❶ 默认为 3 个 TCP 调度器。
❷ 最大支持 10 个调度器。
(3) 配置 PROD 数据库支持如下要求:
❶ 最少 10 个共享服务进程。
❷ 最多 30 个共享服务进程。

```
-- 1.300 session,100 dedicated,修改 sessions 等参数.
alter system set sessions = 300 scope = spfile;
alter system set shared_server_sessions = 200 scope = spfile;
alter system set processes = 400 scope = spfile;
startup force;

修改调度器等参数.
-- 2.PROD,3tcp dispatchers, max 10 dispatchers
alter system set dispatchers = '(protocol = tcp)(dispatchers = 3)(connection = 100)' scope = both;
alter system set max_dispatchers = 10 scope = both;
修改共享服务模式的相关参数.
-- 3.PROD,10 shared processes,max 30
alter system set shared_servers = 10 scope = both;
alter system set max_shared_servers = 30 scope = both;
```

9.5 客户端网络服务配置

(1) 创建一个客户端的网络配置文件,可以使用本地命名方式和简单连接方式连接到数据库中。
❶ 实例 PROD 的别名为 prod,使用默认的 listener 和专用服务模式连接数据库。
❷ 实例 PROD 的第 2 个别名 prod_s,应使用 LSNR2 和使用共享服务模式连接数据库。
(2) 别名 racdb,应连接至数据库服务 RACDB(稍后创建),使用专用服务模式。

这里的 RACDB 服务将运行在用户的 RAC 集群中。

（3）别名 emrep，应连接至 EMREP 实例（稍后创建），使用专用服务模式。

```
-- 1.sqlnet.ora 修改 sqlnet.ora 配置.
NAMES.DIRECTORY_PATH = (TNSNAMES, EZCONNECT)

-- tnsname.ora 修改连接字符串.
prod =
  (DESCRIPTION =
    (ADDRESS_LIST =
      (ADDRESS = (PROTOCOL = TCP)(HOST = odd)(PORT = 1521))
    )
    (CONNECT_DATA =
      (SERVER = DEDICATED)
      (SERVICE_NAME = prod)
    )
  )

prod_s =
  (DESCRIPTION =
    (ADDRESS_LIST =
      (ADDRESS = (PROTOCOL = TCP)(HOST = odd)(PORT = 1526))
    )
    (CONNECT_DATA =
      (SERVER = SHARED)
      (SERVICE_NAME = prod)
    )
  )

racdb =
  (DESCRIPTION =
    (ADDRESS_LIST =
      (ADDRESS = (PROTOCOL = TCP)(HOST = yingshu)(PORT = 1521))
    )
    (CONNECT_DATA =
      (SERVER = DEDICATED)
      (SERVICE_NAME = racdb)
    )
  )

EMREP =
  (DESCRIPTION =
    (ADDRESS_LIST =
      (ADDRESS = (PROTOCOL = TCP)(HOST = odd)(PORT = 1521))
    )
    (CONNECT_DATA =
      (SERVER = DEDICATED)
      (SERVICE_NAME = emrep)
    )
  )
```

```
###########################
-- rac sample             //RAC 的连接字符串
ODSDB =
  (DESCRIPTION =
    (ADDRESS = (PROTOCOL = TCP)(HOST = odsdb1-vip)(PORT = 1521))
    (ADDRESS = (PROTOCOL = TCP)(HOST = odsdb2-vip)(PORT = 1521))
    (LOAD_BALANCE = yes)
    (FAILOVER = ON)
    (CONNECT_DATA =
      (SERVER = DEDICATED)
      (SERVICE_NAME = odsdb)
    )
  )
```

9.6 表空间的创建和配置

(1) 创建一组临时表空间,含有两个临时表空间,使其支持并发创建大索引,分析表等大操作,满足如下要求:

❶ 临时表空间的组名为 TEMP_GRP,包含两个临时表空间分别为 TEMP1 和 TMEP2。

❷ 设置 TEMP_GRP 为所有用户默认的临时表空间。

(2) 创建一个永久表空间,存储测试数据,按照如下要求创建:

❶ 表空间名为 EXAMPLE。

❷ 初始的数据文件为 400MB,可以增长至 4TB。

❸ 初始的区大小为 1MB。

❹ 下一个区的大小为 1MB。

(3) 创建一个永久表空间存储索引,按照如下要求创建:

❶ 表空间名为 INDX。

❷ 大小为 40MB。

(4) 创建一个永久表空间存储收集的各种 Oracle 工具,按照如下要求创建:

❶ 表空间名为 TOOLS。

❷ 数据文件大小为 48MB。

❸ 初始的区大小为 4MB。

❹ 下一个区的大小为 4MB。

(5) 创建一个默认的永久表空间,按照如下要求创建:

❶ 表空间名为 USERS。

❷ 数据文件大小为 48MB。

❸ 初始区的大小为 4MB。

❹ 下一个区的大小为 4MB。

(6) 创建一个永久表空间,考虑到在线的进程需要大量的并发插入操作,应降低该表空间中表的维护成本,按照如下要求创建:

❶ 表空间名为 OLTP。
❷ 文件大小为 48MB。
❸ 初始区的大小为 2MB。
❹ 下一个区的大小为 2MB。

```
1 创建临时表空间.
create temporary tablespace temp1 tempfile '/u01/app/oracle/oradata/PROD/disk4/temp1_01.dbf'
size 100m tablespace group temp_grp;
create temporary tablespace temp2 tempfile '/u01/app/oracle/oradata/PROD/disk4/temp2_01.dbf'
size 100m tablespace group temp_grp;
alter database default temporary tablespace temp_grp;

2 创建大文件表空间 example
create bigfile tablespace example datafile '/u01/app/oracle/oradata/PROD/disk5/example_01.
dbf' size 400m reuse autoextend on extent management local uniform size 1m;

3 创建索引表空间 indx
create tablespace indx datafile '/u01/app/oracle/oradata/PROD/disk5/indx_01.dbf' size 40m
reuse;

4 创建工具表空间 tools
create tablespace tools datafile '/u01/app/oracle/oradata/PROD/disk5/tools_01.dbf' size 10m
reuse;

5 创建 users 表空间
create tablespace users datafile '/u01/app/oracle/oradata/PROD/disk5/users_01.dbf' size
48m reuse
uniform size 4m;

6 创建 OLTP 表空间
create tablespace oltp datafile '/u01/app/oracle/oradata/PROD/disk5/oltp_01.dbf' size 48m
autoextend on extent management local uniform size 2m segment space management auto;

修改数据库默认表空间
alter database default tablespace tools;

删除表空间命令 tablespace
-- drop tablespace users including contents and datafiles cascade constraints;
```

9.7 日志文件管理

（1）预计有很大的事务量，数据库日志应该按照如下配置。
❶ 至少需要配置 5 组日志。
❷ 每一组日志文件都需要冗余。
❸ 日志文件的大小为 100MB。
❹ 日志文件的位置，应该考虑减少 I/O 压力和降低单块磁盘损坏时的故障风险。
（2）复用控制文件，减少因磁盘损坏导致的控制文件需要恢复的风险。

```
alter database add logfile group 4 ('/u01/app/oracle/oradata/PROD/disk1/redo04.log') size
100m reuse;
alter database add logfile group 5 ('/u01/app/oracle/oradata/PROD/disk1/redo05.log') size
100m reuse;
alter database add logfile member '/u01/app/oracle/oradata/PROD/disk2/redo011.log' to
group 1;
alter database add logfile member '/u01/app/oracle/oradata/PROD/disk2/redo021.log' to
group 2;
alter database add logfile member '/u01/app/oracle/oradata/PROD/disk2/redo031.log' to
group 3;
alter database add logfile member '/u01/app/oracle/oradata/PROD/disk2/redo041.log' to
group 4;
alter database add logfile member '/u01/app/oracle/oradata/PROD/disk2/redo051.log' to
group 5;
```

复用控制文件
停止数据库
shutdown immediate;
将 controlfile 拷贝两份;
修改 spfile 中的控制文件参数,指向两个控制文件.

9.8 创建模式(Schema)

使用 sys 用户,执行脚本文件/home/oracle/scripts/create_bishhr.sql,忽略任何因为连接 OE 所致的创建错误,但不能忽略其他创建错误。

```
创建 hr 方案.
conn /as sysdba
@/home/oracle/scripts/create_bishhr.sql

创建 hr 用户.
create user hr
identified by hr
default tablespace example
quota unlimited on example;
grant connect,resource,dba to hr;
alter user hr quota unlimited on example;


创建 sh 用户.
create user sh
identified by sh
default tablespace example
temporary tablespace temp1
quota unlimited on example;
grant connect,resource,dba to sh;
alter user sh quota unlimited on example;
```

9.9 模式的统计信息和参数文件配置

收集数据库中不同模式的统计信息,使基于代价的优化器选择正确的执行路径。观察参数文件中的每一个参数,使其设置合理。增加附加的参数,使数据库处于最优化的状态,并设置以下参数:utl_file_dir=('/home/oralce','/home/oracle/temp','/home/oracle/scripts')

```
开启自动收集统计信息.
begin
dbms_scheduler.enable('GATHER_STATS_JOB');
end;
/

select count(*) from dba_scheduler_jobs where job_name = 'GATHER_STATS_JOB';

手工执行收集全库的统计信息.
exec dbms_stats.gather_database_stats;

修改 utl_file 包可以操作的目录参数.
alter system set utl_file_dir = '/home/oralce','/home/oracle/temp','/home/oracle/scripts'
scope = spfile;


修改 prod 为归档模式.
alter system set log_archive_dest_1 = '/u01/app/oracle/arch_prod' scope=spfile;
shutdown immediate;
startup mount;
alter database archivelog;
alter database open;
alter system switch logfile;
```

9.10 数据库的备份和高可用

备份数据库,使数据库在任何状况下都可以完全恢复。
打开数据库:

```
备份数据库.
run {
backup full tag 'PROD' database
include current controlfile format '/oracle/bak/orclfull_%d_%T_%s'
plus archivelog format '/oracle/bak/arch_%d_%T_%s' delete all input;
}
```

9.11 创建一个数据库

(1) 在你的数据库服务器中再创建一个数据库。

❶ 数据库名和实例名都为 EMREP。

❷ 在 EMREP 数据库中，标签安全(Label Security)是唯一需要安装的选件。
(2) 在你的管理服务器中安装 Grid Control，安装介质在/oramed 下。
(3) 在你的数据库服务器中部署 Oracle 管理代理(Management Agent)服务。
(4) 创建一个 Grid Control 控制台，超级管理员为 EMADMIN，密码为 EMADMIN。

```
用 DBCA 创建一个 EMREP 数据库,选择 label security 功能.
use dbca create EMREP password EMREP
label security

修改 EMREP 数据库参数.
-- EMREP parameter
alter system set sga_target = 300m scope = spfile;
alter system set job_queue_processes = 10 scope = spfile;
alter system set processes = 300 scope = spfile;
alter system set open_cursors = 1000 scope = spfile;
alter system set session_cached_cursors = 200 scope = spfile;
alter system set aq_tm_processes = 1 scope = spfile;
startup force;

创建 DBMS_SHARED_POOL 包.
@?/rdbms/admin/dbmspool.sql
select object_name, object_type, status, owner
from all_objects
where object_name = 'DBMS_SHARED_POOL';
```

9.12　安装 grid control

```
groupadd dba                                    //建用户组
groupadd oinstall
useradd -g oinstall -G dba oracle
passwd oracle

mkdir -p /u01/app/oracle/product/10.2.0         //建安装用的文件夹
chmod -R 755 /u01
chown -R oracle.dba /u01

vi /etc/sysctl.conf
kernel.shmall = 2097152
kernel.shmmax = 2147483648
kernel.shmmni = 4096
# semaphores: semmsl, semmns, semopm, semmni
kernel.sem = 1010 129280 1010 128
fs.file-max = 327679
net.ipv4.ip_local_port_range = 1024 65000
net.core.rmem_default = 262144
net.core.rmem_max = 262144
net.core.wmem_default = 262144
```

```
net.core.wmem_max = 262144

vi /etc/security/limits.conf
oracle soft nproc 16384
oracle hard nproc 16384
oracle soft nofile 327600
oracle hard nofile 327600

vi /etc/pam.d/login
session required /lib/security/pam_limits.so

-- secure oms
./opmnctl stopall
./emctl secure oms

---- secure agent
./emctl stop agent
./emctl secure agent
./emctl start agent
./emctl upload agent
./emctl secure status agent
./emctl status agent

-- opmnctl oms
./opmnctl stopall
./opmnctl startall

-- emctl oms
./emctl start oms
./emctl stop oms
./emctl status oms

-- opmnctl web cache
./opmnctl startproc ias-component=WebCache
./opmnctl stopproc ias-component=WebCache
./opmnctl status

-- emctl ias
./emctl start iasconsole
./emctl stop iasconsole
./emctl status iasconsole

-- emctl dbconsole
./emctl start dbconsole
./emctl stop dbconsole

-- start grid control                //开启 grid control
./lsnrctl start
sqlplus "/as sysdba"
startup
```

```
./emctl start oms
./opmnctl startproc ias-component=WebCache
./emctl start agent
./emctl start iasconsole

-- shutdown grid control
./emctl stop oms
./emctl stop iasconsole
./opmnctl stopall
./emctl stop agent
sqlplus "/as sysdba"
shutdown immediate
./lsnrctl stop

-- gridcontrol configure
alter user dbsnmp identified by dbsnmp account unlock;

-- create user EMADMIN passwd EMADMIN
SETUP -- Administrator                    //使用 gridcontrol 创建用户

-- agentDownload
1. # ln -s /usr/bin/wget /usr/local/bin/wget -- root
2. $ export PATH=$ORACLE_HOME/jdk/bin:$PATH -- oracle
3. http://even:4889/agent_download/10.2.0.1.0/linux
4. $ chmod 755 agentDownload.linux
   $ ./agentDownload.linux -b /u01/app/oracle/product

-- gridcontrol
1. sga_target=280m
2. aq_tm_processes>0
3. session_cached_cursors>200
4. dispatchers is nothing
5. /rdbms/admin/dbmspool.sql
6. mkdir

--
cd /oracle/OracleHomes/agent10g/sysman/emd/upload
rm -rf *
agentca -f
```

9.13 使用 Grid Control

(1) 使用 Grid Control，修改数据库 PROD 的参数 pga_aggregate_target 至 500MB，然后重启数据库。

(2) 使用 Grid Control 配置实例，使其可以在出现实例故障的情况下，5min 之内启动成功。

(3) 在数据库 PROD 中，配置一个警告阈值，当 system 表空间利用率达 90% 时警告，

95%时严重警告。

(4) 设置事件通知,将事件信息发至邮箱 dba@ocm.com。

(5) 使用 Grid Control,在 PROD 数据库中创建一个新的表空间名为 REGISTRATION。

❶ 包含一个数据文件,数据文件的大小为 90MB。

❷ 如果空间不够用,这个数据文件可以增长至 120MB。

❸ 设置这个表空间利用最优的块空间管理。

```
-- grid console,prod 将 sga 修改为 500m.
-- 1.use gc to pga 500m,startup

-- 2.fast_start_mttr_target to 5 min 300 sec        //设置 grid control 参数

-- 3.system 90 % 95critical                          //设置表空间报警门限
database -- prod -- administration -- tablespaces -- system -- edit -- thresholds
specify thresholds 95 % critical ...

-- 4.dba@ocm.com                                    //设置警告日志发向哪个邮件
setup -- notification methods --
outgoing Mail(SMTP) Server: mail.ocm.com
identify sender as :oracle
sender's e-mail address :dba@ocm.com

-- 5.PROD,create tbs REGISTRATION                   //创建表空间
database -- prod -- administration -- tablespace -- create
name:REGISTRATION -- datafile -- add --
```

9.14 实现调度器(Schedules)和定时任务(Jobs)

(1) 使用 Grid Control,在数据库 PROD 中创建一个 Schedules。

❶ 这个调度器的名字叫 DAILYREBUILD。

❷ 这个调度器每天下午 2 点执行。

(2) 在数据库 PROD 中,创建一个程序(Program),名为 EMP_IND_REBUILD,这个程序将重建表 HR.EMPLOYEES 上面所有的索引。

(3) 利用调度器 DAILYREBUILD 和 SYSTEM_PLAN 资源管理计划,创建一个时间窗口(Window)。

(4) 使用调度器 DAILYREBUILD 和程序 EMP_IND_REBUILD 创建一个名为 REBUILD_JOB 的定时任务(Job)。

```
创建一个调度器.
-- 1.gc,prod create schedule

begin
sys.dbms_scheduler.create_schedule(
repeat_interval => 'FREQ=DAILY;BYHOUR=14;BYMINUTE=0;BYSECOND=0',
start_date => systimestamp at time zone 'America/New_York',
```

```
        schedule_name => '"SYS"."DAILYREBUILD"');
end;
```

创建一个程序，让这个程序重建一个表中的索引.
```
-- 2.create program in prod EMP_IND_REBUILD rebuilds all indexes on the hr.EMPLOYEES
select index_name from user_indexes where table_name = 'EMPLOYEES';

begin
dbms_scheduler.create_program(
program_name =>'"HR"."EMP_IND_REBUILD"',
program_action =>'alter index emp_email_uk rebuild online;
alter index emp_emp_id_pk rebuild online;
alter index emp_department_ix rebuild online;
alter index emp_job_ix rebuild online;
alter index emp_manager_ix rebuild online;
alter index emp_name_ix rebuild online;
',
program_type =>'PLSQL_BLOCK',
number_of_arguments =>0,
comments =>'',
enabled => true);
end;

-- 3.create window use dailyrebuild and system_plan
```
创建一个时间窗口.
```
begin
dbms_scheduler.create_window(
window_name =>'"DAILYREBUILD WINDOW"',
resource_plan =>'SYSTEM_PLAN',
start_date => systimestamp at time zone 'America/New_York',
duration => numtodsinterval(60, 'MINUTE'),
repeat_interval => null,
end_date => null,
window_priority =>'LOW',
comments =>'');
end;

-- 4.create a job REBUILD_JOB use DAILYREBUILD schedule EMP_IND_REBUILD
```
创建一个 job.
```
begin
sys.dbms_scheduler.create_job(
job_name => '"HR"."REBUILD_JOB"',
program_name => '"HR"."EMP_IND_REBUILD"',
schedule_name => '"SYS"."DAILYREBUILD"',
job_class => '"DEFAULT_JOB_CLASS"',
auto_drop => FALSE,
enabled => TRUE);
end;
```

9.15 创建一个 RMAN 恢复目录（Catalog）

（1）在数据库 EMREP 中创建一个表空间名为 RC_DATA。
❶ 此表空间为本地管理表空间。
❷ 此表空间包含一个大小为 100MB 的数据文件。
（2）在数据库 EMERP 中，创建一个用户名为 RC_ADMIN，密码也是 RC_ADMIN 的用户。
❶ 这个用户的默认表空间是 RC_DATA。
❷ 授予此用户管理恢复目录的权限。
（3）创建一个恢复目录。
❶ 在数据库 EMREP 中，创建一个恢复目录，所有者是 RC_ADMIN。
❷ 将数据库 PROD 注册到这个恢复目录中。

```
-- 1.create tbs RC_DATA in EMREP.创建一个表空间在 EMREP 库中.
create tablespace rc_data datafile '/u01/app/oracle/oradata/EMREP/rc_data_01.dbf' size 100m;
创建一个用户.
-- 2.create user RC_ADMIN passwd RC_ADMIN EMREP
default tbs rc_data,grant recv cata owner
create user RC_ADMIN
identified by RC_ADMIN
default tablespace rc_data
quota unlimited on rc_data;
grant connect,resource,recovery_catalog_owner to rc_admin;

-- 3.create recovery catalog 创建一个 recovery catalog
rman
connect catalog rc_admin/rc_admin@EMREP
create catalog tablespace rc_data;

rman target sys/oracle@PROD catalog RC_ADMIN/RC_ADMIN@EMREP
register database;                //将数据库注册到 catalog 中去
-- resync catalog;
```

9.16 使用 RMAN

（1）在数据库 PROD 中配置 RMAN 选项
❶ 开启最优备份模式。
❷ 设置默认通道写目录为/home/oracle/backup，必须建好此目录。
❸ 开启控制文件自动备份，将备份路径设置为/home/oracle/backup/control 目录。
❹ 配置保留策略为 7 天。
（2）执行一个备份
❶ 使用压缩功能，使用默认通道执行一个数据库备份。

❷ 包含所有的数据文件。
❸ 包含当前的控制文件和 spfile。
❹ 包含所有的归档日志文件,并将已备份的归档日志删除。

```
CONFIGURE RMAN OPTIONS prod 配置 RMAN 参数.
-- mkdir backup
configure backup optimization on;
configure channel device type disk format '/home/oracle/backup/DB_%U';
configure controlfile autobackup on;
configure controlfile autobackup format for device type disk to '/home/oracle/backup/control/cf_%F';
configure retention policy to recovery window of 7 days;
configure default device type to disk;
configure device type disk backup type to compressed backupset;


-- mkdir 创建备份目录.
/home/oracle/backup
/home/oracle/backup/control

备份数据库.
rman target sys/oracle@PROD catalog RC_ADMIN/RC_ADMIN@EMREP

-- backup all
run {
backup database include current controlfile;
backup archivelog all delete all input;
}
恢复表空间 system
-- recover tablespace.
rman target sys/oracle@PROD catalog RC_ADMIN/RC_ADMIN@EMREP
startup mount;
restore tablespace system;
recover tablespace system;
alter database open;
```

9.17 回闪数据库

打开数据库的回闪功能。
1. 设置回闪恢复区的大小为 4GB。
2. 将回闪目录设置为/home/oracle/flash,必须创建这个目录。
保持数据库处于打开状态。
注:使用回闪功能,回闪一张表。实际在生产中,可以不用回闪功能。

```
-- archivelog mode
-- mkdir -p /home/oracle/flash
-- set db_recovery_file_dest                    //设置回闪目录和回闪日志大小
```

```sql
alter system set db_recovery_file_dest_size = 4g scope = both;
alter system set db_recovery_file_dest = '/home/oracle/flash' scope = spfile;

-- set flashback back                          //开启回闪模式
shutdown immediate;
startup mount;
-- alter database archivelog;
alter database flashback on;
alter database open;

-- grant                                       //授回闪权限
grant flashback any table to hr;
grant flashback on scott.emp to fuli;

flashback database to before point 'before_upgrade';
flashback database to scn 202381

-- show background processce
ps -ef|grep rvw

-- show current scn                            //获取当前的 SCN 号
select name,current_scn,flashback_on from v$database;
CURRENT_SCN
-----------
     661911

-- show current time
select to_char(sysdate,'yy-mm-dd hh24:mi:ss') time from dual;
TIME
------------------
15-11-09 03:37:49

-- set parameter retention
alter system set db_flashback_retention_target = 1440 scope = both;

-- drop a table
drop table employees cascade constraints;
commit;

-- flashback database
shutdown immediate;                            //基于时间点的回闪,基于 SCN 的回闪
startup mount;
flashback database to timestamp to_timestamp('15-11-09 03:37:49','yy-mm-dd hh24:mi:ss');
flashback database to scn 661911;
alter database open resetlogs;

-- some view                                   //和回闪相关的视图
v$database
```

```sql
v$flashback_database_log
V$flashback_database_stat

-- recyclebin                                    //设置回收站
show parameter recycle;
alter system set recyclebin = off;
alter session set recyclebin = off;
drop table name purge;

-- flashback drop                                //从回收站中找出删除的表
drop table employees cascade constraints;
show recyclebin;
select original_name,object_name from recyclebin;
flashback table employees to before drop;
flashback table employees to before drop rename to b;
create table employees as select * from b;

-- as of timestamp
select * from employees as of timestamp sysdate - 5/1440
select * from employees as of timestamp to_timestamp('2015 - 11 - 09 04:32:58','yyyy - mm - dd hh24:mi:ss');
insert into employees select * from employees as of timestamp to_timestamp('2015 - 11 - 09 04:32:57','yyyy - mm - dd hh24:mi:ss');

-- as of scn
select dbms_flashback.get_system_change_number from dual;
select current_scn from v$database;
delete from employees;
commit;
select * from employees as of scn 664016;
insert into employees select * from employees as of scn 664016;

-- time to scn
desc sys.smon_scn_time
select scn from sys.smon_scn_time

-- flashback table
select row_movement from user_tables where table_name = 'EMPLOYEES';

-- need table movement
select current_scn from v$database;
alter session set nls_date_format = "yyyy - mm - dd hh24:mi:ss";
select sysdate from dual;
alter table employees enable row movement;
flashback table employees to timestamp to_timestamp('2015 - 04 - 09 05:32:18','yyyy - mm - dd hh24:mi:ss');
flashback table employees to scn 665364;
```

9.18 实体化视图

快速刷新的实体化视图。

使用 sh 用户创建一个名为 PROD_MV 的快速刷新的实体化视图，查询语句在 mview1.txt 文件中。

```
-- 1.fast refreshable materialized view
create materialized view log on departments with sequence, rowid
(DEPARTMENT_ID, DEPARTMENT_NAME, MANAGER_ID, LOCATION_ID) including new values;

create materialized view mv_fast refresh fast as
select department_id, manager_id from departments group by department_id, manager_id;

create materialized view mv_fact refresh fast as
select a,b from t group by a,b;

prod.hr.EMPLOYEES
conn hr/hr
create materialized view mv_fact refresh fast as
select f.rowid f_rowid, a.rowid a_rowid, b.rowid b_rowid, f.id,
a.name a_name, b.name b_name, num
from fact f, dim_a a, dim_b b
where f.aid = a.id
and f.bid = b.id;

创建实体化视图
hr.employees
create materialized view log on employees with rowid
(EMPLOYEE_ID, FIRST_NAME, LAST_NAME, PHONE_NUMBER) including new values;

create materialized view employees_mv
refresh fast
as select employee_id
from employees group by employee_id;
```

9.19 手工刷新实体化视图

创建一个可以更新的实体化视图。

在 PROD 数据库中，使用 HR.EMPLOYEES 表创建一个名为 EMP_UPD_MV 的可以更新的实体化视图。

```
-- show view
select * from dba_db_links;

-- create link
create public database link l1 connect to system identified by oracle using 'prod';
```

```sql
-- test link
select employee_id,first_name,last_name,phone_number from hr.employees@l1;

-- create link in EMREP
drop database link remote;
drop public database link remote;

-- create emrep EMP_UPD_MV
create materialized view emp_upd_mv tablespace "USERS"
refresh force on demand for update enable query rewrite as select employee_id,first_name,last_name,phone_number from hr.employees@l1;

create materialized view emp_upd_mv tablespace "USERS"
for update enable query rewrite as select employee_id,first_name,last_name,phone_number from hr.employees@l1;

-- manual refresh
exec dbms_mview.refresh('view_test_2','F');

drop materialized view log                        //删除实体化视图日志
drop materialized view log on table1;             //删除某张表上的实体化视图
drop materialized view emp_upd_mv;                //删除实体化视图

create materialized view log on table1
tablespace ts_data
with rowid;
```

9.20 外部表的使用（Oracle_Loader External Tables）

（1）在脚本目录中找到 prod_master.dat 和 prod_master.ctl 两个文件。在 RPOD 数据库中，在 SH 用户下使用这两个文件创建一个名为 PROD_MASTER 的外部表。

```sql
-- prod_master.dat and prod_master.ctl create PROD_MASTER sh schema prod
-- pdf in Utilities Part III External Tables
-- create directory
connect sh/sh
create or replace directory l1 as '/home/oracle/scripts';
conn /as sysdba
grant write,read on directory l1 to sh;

-- show directory
col directory_path for a30
col owner for a10
select * from dba_directories;
```

```
-- create external table
conn sh/sh
drop table prod_master;

    //编辑如下文本内容,将在数据库中由内部表读出
-- vi /home/oracle/scripts/prod_master.dat
7369,SMITH,CLERK,7902,17-DEC-80,100,0,20
7499,ALLEN,SALESMAN,7698,20-FEB-81,250,0,30
7521,WARD,SALESMAN,7698,22-FEB-81,450,0,30
7566,JONES,MANAGER,7839,02-APR-81,1150,0,20

-- create prod_mater external organization
    conn sh/sh
    drop table prod_master;
    create table prod_master
      (emp_id number(4),
       ename varchar2(12),
       job varchar2(12),
       mgr_id number(4),
       hiredate date,
       salary number(8),
       comm number(8),
       dept_id number(2))
    organization external
    (type oracle_loader
    default directory ll
    access parameters(records delimited by newline
                     fields terminated by ','
                     missing field values are null
(
    emp_id,
    ename,
    job,
    mgr_id,
    hiredate char date_format date mask 'dd-mon-yyyy',
    salary,
    comm,
    dept_id ))
    location('prod_master.dat'))
  reject limit unlimited;

-- query table
select * from prod_master;
```

数据泵外部表 Oracle_Datapump External Table

（2）在数据库 PROD 中使用 SH 用户,创建一个名为 COUNTRIES_EXT 的外部表。基础表为 SH.COUNTRIES,此外部表包含的列为 COUNTRY_ID、COUNTRY_NAME 和 COUNTRY_REGION。

```sql
-- 1. prod sh external COUNTRIES_EXT
conn sh/sh@prod
select * from dba_directories;
create table countries_ext organization external
(
  type oracle_datapump
  default directory ll
  location ('COUNTRIES_EXT.dmp')
)
as select country_id,country_name,country_region from countries;

select count(*) from countries_ext;
select * from dba_external_locations;

-- 2. EMREP db system COUNTRIES_EXT use COUNTRIES_EXT.dmp
conn system/oracle@emrep

-- create directory
connect system/oracle
create or replace directory ll as '/home/oracle/scripts';
conn /as sysdba
grant write,read on directory ll to system;

select * from dba_directories;

-- create COUNTRIES_EXT
conn system/oracle
create table countries_ext
  (
      country_id      char(2),
      country_name    varchar2(40),
      country_region  varchar2(40)
  )
organization external
  (
      type oracle_datapump
      default directory LL
      location('COUNTRIES_EXT.dmp')
  );

select * from countries_ext;
drop table countries;
```

9.21 传输表空间(Transportable Tablespace)

(1) 使用导入导出工具将文件 sst.dmp 中的所有对象导入到数据库 PROD 的 OLTP_USER 用户中,此文件是由 SST 用户导出的。

(2) 使用表空间传输,将表空间 OLTP 由 PROD 数据库传至 EMREP 数据库。完成这个任务之后,OLTP 表空间应在两个数据库中都可以读写,并且 PROD 数据库中 OLTP_USER 用户下的所有对象在 EMERP 数据库中都可以正常使用。

```
-- create user oltp_user
drop user oltp_user;
create user oltp_user
identified by oltp_user
default tablespace oltp
temporary tablespace temp1
quota unlimited on oltp;
grant connect,dba,resource to oltp_user;

-- imp or impdp sst.dmp to user OLPT_USER fromuser sst
impdp system/oracle file = sst.dmp fromuser = sst touser = OLTP_USER ignore = y

-- create table
conn oltp_user/oltp_user
create table trans(id number) tablespace oltp;
insert into trans values (1);
commit;

-- full dependency check
conn /as sysdba
exec dbms_tts.transport_set_check('oltp',true);
select * from transport_set_violations;

-- create directory
create directory dir1 as '/oracle/trans';
grant read,write on directory dir1 to system;

-- 2.tts OLTP PROD TO EMREP
-- sample 1
-- expdp
ss
alter tablespace oltp read only;
cp /u01/app/oracle/oradata/PROD/disk3/oltp01.dbf /u01/app/oracle/oradata/EMREP/oltp_01.dbf
expdp system/oracle dumpfile = exptbsoltp.dmp directory = dir1 transport_tablespaces = OLTP transport_full_check = y
expdp system/oracle dumpfile = exptbsoltp.dmp transport_tablespaces = oltp transport_full_check = y

-- impdp
export ORACLE_SID = EMREP
ss
#create directory workdir as '/oracle/trans';
#grant read,write on directory workdir to system;
#drop tablespace oltp including contents and datafiles cascade constraints;
-- create user oltp_user
```

```
create user oltp_user identified by oltp_user;
grant connect,dba,resource to oltp_user;
impdp system/oracle dumpfile = exptbsoltp.dmp transport_datafiles = '/u01/app/oracle/oradata/
EMREP/oltp_01.dbf' remap_schema = (oltp_user:oltp_user)
alter user oltp_user default tablespace oltp;
alter tablespace oltp read write;
-- prod db
alter tablespace oltp read write;
# conn oltp_user/oltp_user
# imp fromuser = sys touser = oltp_user file = expttsoltp.dmp transport_tablespace = y datafiles
 = ('/u01/app/oracle/oradata/EMREP/oltp_01.dbf')
```

9.22 创建一个附加的数据缓冲区

在 PROD 数据库中，创建一个附加的数据缓冲区，块大小是 16KB，确保这个缓冲区处于可用状态。

```
alter system set db_16k_cache_size = 16m scope = both;
```

9.23 创建大文件表空间

在 PROD 数据库中，新创建一个名为 LOB_DATA 的大对象表空间，用于存储大对象和大对象索引，按下面的要求创建：
(1) 创建两个数据文件，分别存放在不同的地方。
(2) 每个数据文件大小为 64MB。
(3) 块大小为 16KB。
(4) 确认区管理模式处于最佳状态。

```
-- lob PROD
create tablespace lob_data datafile '/u01/app/oracle/oradata/PROD/disk4/lob_data_01.dbf'
size 64m uniform size 4m blocksize 16k;
alter tablespace lob_data add datafile '/u01/app/oracle/oradata/PROD/disk5/lob_data_02.dbf'
size 64m;
```

9.24 管理用户数据

(1) 按下面的要求，在 PROD 数据库中，使用 HR 用户创建一个新表。
❶ 表名为 MAGAZINE_ARTICLES
❷ 表空间为 USERS
❸ 列名为
AUTHOR   VARCHAR2(30)
ARTICLE_NAME   VARCHAR2(50)

ARTICLE_DATE   DATE
ARTICLE_DATA   CLOB

（1）表空间 LOB_DATA 是 16KB 的大块，初始的区和下一个区都是 2MB。

（2）使用 nocache 和 disable storage in row 选项。

❹ 将 exp_mag.dmp 文件中的数据导入到表 HR.MAGAZINE_ARTICLES 中去。

（2）在 PROD 数据库中，使用 hr 用户按照下面的要求创建一个新表。

❶ 表名为 ORACLE9I_REFERENCES。

❷ 存储在表空间 USERS 中。

❸ 表结构为：

ORACLE9I_ARTICLE ROWID,
INSERT_TIME TIMESTAMP WITH LOCAL TIME ZONE

（3）将表 HR.MAGAZINE_ARTICLES 中所有符合条件的数据相关的列插入到表 ORACLE9I_REFERENCES 中去，条件是包含 3 个或更多的 Oracle 9i 字符。

```
-- 1.MAGEZINE_ARTICLES

alter user hr quota unlimited on lob_data;

-- create tb MAGAZINE_ARTICLES
conn hr/hr
drop table magazine_articles;
create table magazine_articles
(
author            varchar2(30),
article_name      varchar2(50),
article_date      date,
article_data      clob
)
lob ("ARTICLE_DATA") store as asdf
(tablespace "LOB_DATA" disable storage in row chunk 16384 nocache storage(initial 2m next 2m
pctincrease 0))
tablespace users;


-- 1.4 imp exp_mag.dmp
imp hr/hr file = exp_mag.dmp table = magazine_articles


-- 2.user HR PROD db tbname ORACLE9I_REFERENCES
conn hr/hr
drop table oracle9i_references;
create table oracle9i_references
( oracle9i_article rowid,
  insert_time timestamp with local time zone
)tablespace users;


-- 3.hr.magazine_articles Oracle9i' insert into ORACLE9I_REFERENCES
```

```
insert into hr.oracle9i_references (oracle9i_article,insert_time) select
rowid,current_timestamp from hr.magazine_articles where
article_data like 'Oracle9i%Oracle9i%Oracle9i%';
```

9.25 分区表

(1) 按下面的要求,在数据库 PROD 中,创建 5 个新的表空间。

❶ 表空间名为 DATA01、DATA02、DATA03、DATA04 和 DATA05。

❷ 指定数据文件分布在不同的磁盘设备上。

❸ 每一个数据文件大小为 256MB。

❹ 区统一大小为 4MB。

❺ 块的大小为 16KB。

注:这些表空间用于存储分区表。

(2) 按照下面的要求,在 PROD 数据库中使用 sh 用户,创建一个分区表,表名为 SALES_HISTORY。列名和表的定义与表 OLTP_USER.SALES 一样。

❶ 分区 P1 包含 1998 年的数据存储在表空间 DATA01 中。

❷ 分区 P2 包含 1999 年的数据存储在表空间 DATA02 中。

❸ 分区 P3 包含 2000 年的数据存储在表空间 DATA03 中。

❹ 分区 P4 包含 2001 年的数据存储在表空间 DATA04 中。

❺ 分区 P5 包含 2002 年的数据存储在表空间 DATA05 中。

(3) 执行目录/home/oracle/scripts 下的脚本文件 populate_sales_hist.sql,将数据插入到 SALES_HISTORY 表中。

(4) 按下面的要求,在 PROD 数据库中,使用 sh 用户,在表 SALES_HISTORY 上创建一个唯一索引名为 SALES_HISTORY_PK 的文件,将索引分为 4 个分区,使每个分区包含数量相当的数据行数。

❶ 索引包含 ORDERID 列。

❷ 索引存储在表空间 INDX 中。

❸ 创建索引使用的并行度为 4。

```
1. -- create 5 tbs in PROD
alter system set db_4k_cache_size = 4m scope = both;
alter system set db_16k_cache_size = 16m scope = both;

create tablespace data01 datafile '/u01/app/oracle/oradata/PROD/disk1/data01_01.dbf' size
256m uniform size 4m blocksize 16k;
create tablespace data02 datafile '/u01/app/oracle/oradata/PROD/disk2/data02_01.dbf' size
256m uniform size 4m blocksize 16k;
create tablespace data03 datafile '/u01/app/oracle/oradata/PROD/disk3/data03_01.dbf' size
256m uniform size 4m blocksize 16k;
create tablespace data04 datafile '/u01/app/oracle/oradata/PROD/disk4/data04_01.dbf' size
256m uniform size 4m blocksize 16k;
```

```sql
create tablespace data05 datafile '/u01/app/oracle/oradata/PROD/disk5/data05_01.dbf' size
256m uniform size 4m blocksize 16k;

--2.create part tb SALES_HISTORY sh user range part
conn sh/sh
drop table sales_history;
create table sales_history
(orderid number(5),
salesman_name varchar2(30),
sales_amount number(10),
sales_date date)
compress
partition by range(sales_date)
(partition sales_1998 values less than(to_date('01/01/1999','dd/mm/yyyy')) tablespace
data01,
Partition sales_1999 values less than(to_date('01/01/2000','dd/mm/yyyy')) tablespace data02,
Partition sales_2000 values less than(to_date('01/01/2001','dd/mm/yyyy')) tablespace data03,
Partition sales_2001 values less than(to_date('01/01/2002','dd/mm/yyyy')) tablespace data04,
Partition sales_2002 values less than(to_date('01/01/2003','dd/mm/yyyy')) tablespace
data05);


--3.ran /home/oracle/scripts/populate_sales_hist.sql insert into SALES_HISTORY
@/home/oracle/scripts/populate_sales_hist.sql

--4.queue index sales_history_pk hr prod tb sales_history
drop index sales_history_pk;
create unique index sales_history_pk on sales_history("ORDERID")
parallel 4
global partition by hash ("ORDERID")
partitions 4 store in (indx,indx,indx,indx);

create unique index sales_history_pk on sales_history("ORDERID")
parallel 4
global partition by hash("ORDERID")
(
partition "SALES_HISTORY_PK_P1" tablespace "INDX",
partition "SALES_HISTORY_PK_P2" tablespace "INDX",
partition "SALES_HISTORY_PK_P3" tablespace "INDX",
partition "SALES_HISTORY_PK_P4" tablespace "INDX"
);

--5.SALES_HISTORY_DATE_IDX
drop index local_prefixed;
create index sales_history_date_idx on sales_history(sales_date) local;
create index local_noprefixed on sales_history(sales_date) local;

SQL> create index partitioned_index on student_history (graduation_date)
global
partition by range(graduation_date)
```

```sql
(
   partition p_1997 values less than
     (to_date('1997-01-01','yyyy-mm-dd')) tablespace tbs1,
   partition p_1998 values less than
     (to_date('1998-01-01','yyyy-mm-dd')) tablespace tbs2,
   partition p_1999 values less than
     (to_date('1999-01-01','yyyy-mm-dd')) tablespace tbs3,
   partition PARTITIONP_2000 values less than
     (to_date('2001-01-01','yyyy-mm-dd')) tablespace tbs4,
   partition p_max values less than (maxvalue) tablespace tbs4
   );

create index sales_history_date_idx on sales_history(sales_date)
(partition ex1 tablespace indx,
 partition ex2 tablespace indx,
 partition ex3 tablespace indx,
 partition ex4 tablespace indx,
 partition ex5 tablespace indx,
) local;

create index month_ix on sales(sales_month)
    global partition by range(sales_month)
       (partition pm1_ix values less than (2)
         partition pm2_ix values less than (3)
         partition pm3_ix values less than (4)
         partition pm12_ix values less than (maxvalue));

alter index scuba
    rebuild subpartition bcd_types
    tablespace tbs23 parallel (degree 2);


-- hash partition

SQL> create table simple
       ( idx number, txt varchar2(20) )
    partition by hash ( idx )
       partitions 4
       store in ( hist_tab01, hist_tab02 ) ;

-- list partition
create table simple8
      (idx number, txt varchar2(20))
    partition by list (txt)
    ( partition s_top values ( 'HIGH', 'MED' )
        tablespace hist_tab01
    , partition s_bot values( 'LOW', 'MID',NULL )
        tablespace hist_tab02
    ) ;
```

```
-- list
drop table sales_history;
create table sales_history
(
orderid            number,
deptname           varchar2(20),
quarterly_sales    number(10,2),
state              varchar2(10)
)
partition by list (state)
(
partition p1 values (1998) tablespace data01,
partition p2 values (1999) tablespace data02,
partition p3 values (2000) tablespace data03,
partition p4 values (2001) tablespace data04,
partition p5 values (2002) tablespace data05
);
```

9.26 细粒度审计

```
-- add
begin
dbms_fga.add_policy
(object_schema => 'hr',
object_name =>'employees',
policy_name => 'emp_sal',
audit_condition => 'salary > 10000',
audit_column =>'salary',
handler_schema => NULL,
handler_module => NULL,
enable => TRUE,
statement_types => 'SELECT,UPDATE',
audit_trail => DBMS_FGA.DB + DBMS_FGA.EXTENDED,
audit_column_opts => DBMS_FGA.ALL_COLUMNS);
end;
/

-- drop
begin
dbms_fga.drop_policy (
object_schema => 'HR',
object_name => 'EMPLOYEES',
policy_name => 'EMP_SAL'
);
end;
/
```

```sql
-- add
begin
dbms_fga.add_policy(
object_schema =>'HR',
object_name =>'EMPLOYEES',
audit_column =>'SALARY',
audit_condition => 'SALARY >= 10000',
policy_name =>'EMP_ACCESS',
enable => TRUE,
statement_types =>'SELECT,UPDATE'
);
end;
/


-- drop
begin
dbms_fga.drop_policy(
object_schema => 'HR',
object_name => 'EMPLOYEES',
policy_name => 'EMP_ACCESS'
);
end;
/



-- show result
conn hr/hr
select * from employees;
conn /as sysdba
select timestamp, db_user, os_user, object_schema, object_name, sql_text from dba_fga_audit_trail;

-- disable
begin
dbms_fga.enable_policy(
object_schema => 'HR',
object_name => 'EMPLOYEES',
policy_name => 'EMP_ACCESS',
enable => FALSE
);
end;
/

-- enable
begin
dbms_fga.enable_policy(
object_schema =>'HR',
object_name =>'EMPLOYEES',
policy_name =>'EMP_ACCESS');
end;
```

```
/

-- some view
dba_audit_policies
fga_log$  dba_fga_audit_trail
```

9.27 配置资源管理(使用 Grid Control 操作)

按要求配置资源管理。
(1) 将用户 sh 设置为资源管理的超级用户。
(2) 创建两个资源管理消费组,分别为 OLTP 和 DSS。
(3) 使用下面的指令,创建一个计划名为 WEEKDAYS:

❶ 在 OLTP 组,活动的会话数不允许超过 20 个,超过 20 个的活动会话将等待资源,等待超过 60s 将被中断。

❷ 在 DSS 组,最大的活动会话数不超过 5 个,超过 5 个的活动会话将等待资源,等待 120s 仍无资源将被中断。

❸ 在 OLTP 组,查询语句的执行时间最多为 5s,超过 5s 的查询将自动地被切换到 DSS 组。

❹ 在 OLTP 组,允许使用最大的 undo 大小为 200MB。

❺ 将 CPU 的使用比例按如下百分比分配,OLTP、DSS、OHTER_GROUPS 比例分别为 50%、30%、20%。

❻ 设置 DSS 组的并行度为 20。

❼ 在 OLTP 组中,空闲会话不能锁住 DML 语句超过 60s。

(4) 指定用户 OLTP_USER 到 OLTP 资源组。
(5) 指定用户 sh 到 DSS 资源组。
(6) 设置实例的默认资源计划为 WEEKDAYS。

```
--1.SH as administrator
Administration -- users -- sh -- Edit -- System Privlleges -- EDIT List --
ADMINISTER RESOURCE MANAGER -- Move -- Apply

exec dbms_resource_manager_privs.grant_system_privilege
(grantee_name => 'SH',
privilege_name => 'ADMINISTER_RESOURCE_MANAGER',
admin_option => false);

--2.create group OLTP DSS
Administration -- Consumer Groups -- create -- OLTP,DSS

--4.OLTP_USER default group OLTP
Consumer Groups -- OLTP -- edit -- add -- OLTP_USER
```

```
-- 5.SH DSS
Consumer Groups -- DSS -- edit -- add -- DSS

-- 6.dafualt plan WEEKDAYS
alter system set resource_manager_plan = "WEEKDAYS" scope = both;
```

9.28 管理实例的内存结构

(1) 使用 sys 用户创建一个视图,列出在内存中的大于 50KB 的包、存储过程、触发器和函数。视图的名字叫 LARGE_PROC,并且所有用户都可以通过同义词 LARGE_PROC 来访问它。

(2) 设置你的 SGA 为 512MB,开启自动共享内存管理,然后重启数据库。

(3) 开发人员要求设置 Java Pool 为 200MB。

(4) 设置 PGA 的大小为 150MB。

```
-- 1.by sys create LARGE_PROC than 50k object
conn /as sysdba
create view large_proc as select * from dba_object_size where source_size>51200 and type in (
'PACKAGE','PROCEDURE','TRIGGER','FUNCTION');
grant select on large_proc to public;

-- 2.set max_SGA 512M auto sga admin
alter system set sga_max_size = 512m scope = spfile;
alter system set sga_target = 300m scope = both;
startup force;

-- 3.set java pool to 20m
alter system set java_pool_size = 20m scope = both;

-- 4.set pga 150M
alter system set pga_aggregate_target = 150m scope = both;
```

9.29 管理对象的性能

(1) 我们的应用需要访问 SH 用户下的表 CUSTOMERS 上的 CUST_LAST_NAME 列,现在的问题是由于用户疏忽,提供的用户名存在问题,应用将所有的用户名都改成了大写。分析结果显示,我们在此列上建的普通索引应用用户根本没有使用,创建一个合适的索引,让前面提到的应用可以使用此索引。

(2) 打开表 OLTP_USER.SALES 的索引的监控。

(3) 在 OLTP_USER 用户下,创建两个新表分别为 STUDENTS 和 ATTENDEES。STUDENTS 表包含 3 个列,STUD_ID 是 number 类型和主键,FNAME 和 LNAME 是另外两个列,它的长度最大为 20 个字符。ATTENDEES 表是 STUDENTS 和 CLASSES 两张表的交集表,存在多对多的关系。ATTENDEES 表的外键是从其他表的主键中得来的,况且这个表的主键索引和表本身是一个对象（创建一个索引组织表）。

（4）收集表的统计信息。

（5）建位图索引。

（6）建复合压缩索引。

（7）将包 keep 到内存中。

（8）建议使用绑定变量。

（9）将表空间 CUST_TBS 改为 ASSM 表空间。

（10）使用 outline 固定一个 SQL 语句的执行计划。

```
-- 1. -- function upper
create index customers_index on customers (upper(cust_last_name));

-- 2. monitoring index
select index_name from dba_indexes where table_owner = 'OLTP_USER' and table_name = 'SALES';
alter index index_name monitoring usage;
alter index index_name nomonitoring usage;

-- 3.
alter user oltp_user quota unlimited on oltp;

conn oltp_user/oltp_user
create table students (
stud_id number(19) primary key,
fname varchar2(20),
lname varchar2(20)
);
create table attendees
( stud_id number,
class_id number,
primary key (stud_id, class_id) validate ,
foreign key (stud_id) references students (stud_id) validate ,
foreign key (class_id) references classes (class_id) validate )
organization index;


-- 4. Because of the unevenly distributed
create index emp_rev_ind on employees(department_id) reverse;
alter index dept_location_ix rebuild;
dbms_stats.gather_table_stats(ownname =>'HR',TABNAME =>'EMPLOYEES');

-- 5. Analysis has revealed that the country_id column of the customers
create bitmap index cus_bit on customers(country_id);

-- 6. create an index on the country_id and cust_city in customers
create index cus_indx on customers(country_id,cust_city) compress 1;

-- 7. make certain that the package named standard
exec dbms_shared_pool.keep('STANDARD');
```

```sql
alter system set db_keep_cache_size = 10m scope = spfile;
create table tab6(id int) storage (buffer_pool keep);


-- 8. Analysis reveals that a 3rd party application
alter system set cursor_sharing = similar scope = both;

-- 9.

select segment_name, segment_type from dba_segments where tablespace_name = 'CUST_TBS';
```
然后把查到的表 move 到其他表空间
把索引 rebuild 到其他表空间
```sql
alter table table_name move tablespace_name;
alter index index_name rebuild tablespace tablespace_name;
```
删除表空间 cust_tbs, 重新建 cust_tbs 为 ASSM 表空间;
将表 move 回来, 索引重建到新表空间中。

```sql
-- 10. The following SQL statement is used often
Grant create any outline to sh;
conn sh/sh;
create outline cust_city_q on select cust_email from sh.customers where cust_city = 'DENVER';

-- Unlock outl
alter user outln identified by outln account unlock;

alter system set create_stored_outlines = true;
grant create any outline to user_name;
create outline salaries on select last_name, salary from employees;
alter session set use_stored_outlines = true;

-- ouline signature, sys
connect sys/manager
exec dbms_outln.update_signatures;

-- stop db outline
alter system set use_stored_outlines = special;
alter system set use_stored_outlines = false;

-- disable/enable outline:
alter outline ol_name disable;
alter outline ol_name enable;

-- outline category:
exec dbms_outln.drop_by_cat('category_name');

-- outline
select name, category, owner from dba_outlines;
select * from dba_outline_hints;

select name, category, used, sql_text from user_outlines where category = 'DEMO';
```

9.30 statspack 报告

(1) 安装 statspack 包。

❶ 将 TOOLS 表空间作为 perfstat 用户的默认表空间。

❷ 将 TEMP1 表空间作为 perfstat 用户的默认临时表空间。

(2) 收集一个初始的快照，确认包含时间信息和段统计信息，并为此次收集添加注释"MANUAL"。

(3) 在运行脚本 oltp_workload.sql 期间，设置 statspack 每 15min 收集一次快照，然后去掉这个 Job 停止自动收集。

(4) 在目录 /home/oracle 下，生成一个名为 statspack.lst 的 statspack 报告。

```
--1.1PERFSTAT tbs TOOLS TEMP1
创建 statspack 用的表空间.
alter system set job_queue_processes = 6 scope = both;
alter database datafile '/u01/app/oracle/oradata/prod/disk5/tools_01.dbf' resize 200m;
删除 statspack.
-- drop statspack
@?/rdbms/admin/spdrop.sql
创建 statspack.
-- create statspack
@?/rdbms/admin/spcreate.sql
改变收集快照的级别.
--2.i_ucomment
alter system set timed_statistics = true scope = both;
exec statspack.snap(i_snap_level = >7,i_ucomment = >'MANUAL')
execute statspack.modify_statspack_parameter(i_snap_level = >7,i_ucomment = >'MANUAL')
每 15 分钟收集一次快照.
--3.every 5 for 15min
CP SPAUTO.SQL SPAUTO1.SQL
dbms_job.submit(:jobno, 'statspack.snap(i_ucomment = ''MANUAL'')', sysdate, 'trunc(SYSDATE +
1/288,''MI'')', TRUE, :instno);
用 perfstat 用户.
use perfstat user
conn perfstat/perfstat
@?/rdbms/admin/spauto1.sql
select job,what from user_jobs;
select job,next_date,next_sec from user_jobs where job = '&t';


select * from dba_snapshots;
select snap_id from stats $ snapshot;

-- remove statspack job
conn /as sysdba
exec dbms_job.remove(1)
exec dbms_job.broken(job = >1,broken = >true);
```

```
-- 4.generate statspack.lst in /home/oracle
cd /home/oracle
sqlplus /as sysdba
生成 statspack 报告.
@?/rdbms/admin/spreport.sql
-- drop
delete from stats$snapshot where snap_id <＝166;
```

9.31　安装 RAC

参考第 4 章章节。

```
-- raw 裸设备的划分.
vi /etc/sysconfig/rawdevices
# VOTE
/dev/raw/raw1 /dev/sdd1
# OCR
/dev/raw/raw2 /dev/sdd2
vi /etc/rc.d/rc.local
# ocr
chown root:oinstall /dev/raw/raw1
chmod 660 /dev/raw/raw1
# vote
chown oracle:oinstall /dev/raw/raw2
chmod 644 /dev/raw/raw2


-- 配置裸设备
-- 查看此硬盘多大
fdisk -l /dev/sde
-- 给磁盘分 part.
fdisk /dev/sdb
-- 同步分区命令,分完区之后,在另外一个节点上执行.
# partprobe /dev/sde
-- 编辑裸设备文件(双节点)
vi /etc/sysconfig/rawdevices
加入以下信息
# VOTE
/dev/raw/raw1 /dev/sdd1
# OCR
/dev/raw/raw2 /dev/sdd2
-- 强制更新裸设备属组信息.
vi /etc/rc.d/rc.local
# ocr
chown root:oinstall /dev/raw/raw1
chmod 660 /dev/raw/raw1
# vote
chown oracle:oinstall /dev/raw/raw2
```

```
chmod 644 /dev/raw/raw2

-- 手工建立/dev/raw/raw1
touch /dev/raw/raw1
touch /dev/raw/raw2

-- 配置裸设备
CONFIG_MAX_RAW_DEVS = 256
CONFIG_RAW_DRIVER = Y
-- 使裸设备生效
/etc/init.d/rawdevices restart
-- 保证机器启动的时候裸设备能够加载,这一步很重要
/sbin/chkconfig rawdevices on
./runcluvfy.sh stage - pre crsinst - n rac1,rac2 - verbose
先 vipca 再点 ok
```

9.32 配置 DataGuard

参见第 12 章 DataGuard 章节

第 10 章 数据库的升级和补丁

10.1 版本补丁概况

数据库系统规划好之后,就要安装一个全新的数据库,不能安装一个最初发行的版本,一定要安装最后的稳定版本。购买正版的 Oracle,Oracle 官方给的那个介质就是最初发行的版本,安装完成之后必须打补丁,或者直接从 https://support.oracle.com 上下载最后的版本和补丁安装才行。如果只安装一个 Oracle 10.2.0.1/11.2.0.1/12.1.0.2.0 这样的版本,那 bug 就多了,后患无穷,运行时警告日志的报错肯定是一堆堆的,况且还没办法处理,只能升级,报错太多了。

Oracle 软件没有序列号,从网上免费下载的和购买的正版的光盘介质完全一样,下载后可以用于学习和测试运行。但只能下载最初发行的版本,在 https://support.oracle.com 上下载,是需要账号的,这个账号需要购买授权之后才能申请到,账号是需要花钱的。Oracle 的理念就是最初发行的版本随便用,等大家用得顺手了,开始卖授权,卖服务,有了账号才能升级到稳定的版本。数据库的授权就是一张纸,授权这张纸卖得很贵,让 20% 的大公司大单位给 Oracle 提供 80% 的利润。

刚才说了,千万不要把数据库装成 Oracle 10.2.0.1.0/11.2.0.1.0/12.1.0.2.0 这样的版本,这样的版本 bug 很多,很难长期稳定运行。那么,要装成什么版本呢?

(1) Oracle 9i 要装成 Oracle 9.2.0.8.6;

(2) Oracle 10g 要装成 Oracle 10.2.0.5.12;

(3) Oracle 11g 当前的最新的要装成 Oracle 11.2.0.4.7;

(4) Oracle 12c 要安装 Oracle 12.1.0.2.4。

Oracle 12c 因为目前直接生产的还比较少,等 Oracle 12c 第二版发行之后,Oracle 12c 才能陆续安装生产。Oracle 11g 还有三四年的旺盛期,一般客户在投产了以后,三四年是不会再轻易动它了。Oracle 11g Grid Infrastructure PSU 补丁现在很大,看来 Oracle 11.2.0.4 也不是 Oracle 11g 最后的一个 release 版本,还要更新。

版本 11.2.0.4.7 这几个数字分别代表什么意思呢？下面分拆成 11、2、0、4、7 来说明。

11 是 Major Database Release Number，数据库的主版本号，革命性的更新会反映在这个版本号上，如 8i 到 9i、10g。i 代表 internet，g 代表 grid，是从网络到网格的变化，grid 来源于电网，数据库系统形成一个像电网一样的服务系统，就是有几个电厂损坏或毁灭，也不影响整个电网的运行和提供服务。从 11g 到 12c，c 代表 cloud 云。云是分布式计算，将大的计算分拆成较小的子程序。

2 是 Database Maintenance Release Number，数据库维护版本号，包括一些新功能的引入，大的功能变革，会体现在这个版本上，11g 第 2 版就不支持裸设备了，在第 1 版还支持。

0 代表 Application Server Release Number，反映 Oracle 应用服务的发行版本号。

4 是 Component Specific Release Number，组件特效发行版本号。组件的每次更新都会反映在这个版本上。其中就包含补丁包集合，它影响数据库的稳定性，所以更新这个版本号很重要，最好更新至最后一个发布版本。

7 代表 Platform-Specific Release Number，操作系统平台相关的补丁包或补丁包集合，这个补丁包，也应该更新到最新补丁，它修复了一些特定的 bug，增强了安全性。

2016 年 1 月推出了对 PSU、SPU、Bundle Patch 新的命名规则。新的命名规则为（以 11.2.0.4 为例）11.2.0.4.YYMMDD。此处的 YYMMDD 为 patch（PSU、SPU、Bundle）发布的具体日期，格式为两位年份＋两位月份＋两位日期。

例如对 11.2.0.4 推出的第 9 个 DB PSU（本来应该命名为 11.2.0.4.9），在新的命名规则下，这个 patch 被命名为 11.2.0.4.160119，这也表示这 11.2.0.4 的 PSU 是在 2016 年 1 月 19 日推出的 patch。

Oracle 最新补丁号码名称请参考文档 Quick Reference to Patch Numbers for Database PSU，SPU(CPU)，Bundle Patches and Patchsets（Doc ID 1454618.1）。

10.2 补丁的分类

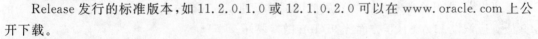

Release 发行的标准版本，如 11.2.0.1.0 或 12.1.0.2.0 可以在 www.oracle.com 上公开下载。

PSR（Patch Set Release）是主版本发布的补丁集，修复了较多的 bug，比如 11.2.0.1 是一个主版本，那么 11.2.0.3 和 11.2.0.4 就是两个 patch set，这些补丁经过了严格的测试，可以放心安全地使用，所以务必安装最新的 patch set。

PSU（Patch Set Updates）为补丁集升级，就是通常说的小补丁。打完之后，它体现在 11.2.0.4.7 的最后一位上。目前最新的就是这个 7。每个季度更新一次，是累积补丁，后一个包含前一个。可以修复一些较严重的问题。最新的 PSU 可以在 support.oracle.com 上搜 PSU 查找相关文档，一般结果都排在最前面，确认最新的 PSU 补丁。这些补丁经过了严格的集成测试，推荐打到最新的 PSU。

CPU（Critical Patch Update）的直接翻译就是"危险的补丁升级"。也是 3 个月更新一次，累积补丁，主要修复安全方面的问题。

One-Off/Interim Patch 是一次性、暂时性补丁，为了解决某一(几)个特定的问题，暂时先打一个小补丁，作为临时的解决办法。一般在测试库上测试通过后，再打到生产库上。

Bundle Patches 补丁包或翻译成 Windows 补丁包，在 Windows 平台上的补丁不叫 PSU，叫 bundle，也是累积补丁，最新的包含之前的补丁。同样建议安装最新的 bundle patch。

10.3 升级前的准备工作

为了防止在升级过程中出现问题，需要有一套回退机制，在紧急状态下，使用这套方案，避免升级失败造成严重损失。最差的结果就是相当于我们没有升级，还是原来的版本，还是原来的数据库状态，不至于数据库无法启动。到了生产的时间，还是不能恢复服务，那问题就严重了。

准备阶段的工作应该包括：
(1) 备份数据库，RMAN 或 expdp 导出。
(2) 备份 CRS 软件，root 用户使用 tar -cvf 命令备份安装目录。
(3) 备份数据库软件，root 用户使用 tar -cvf 命令备份安装目录。
(4) 备份 ASM 盘。asmcmd md_backup 命令。

10.4 版本升级

9i 数据库，必须先安装 9.2.0.1.0，然后再升级至 9.2.0.8.0，然后再打 PSU 或 bundle 补丁至 9.2.0.8.6。10g 也类似，必须先安装 10.2.0.1.0，然后升级至 10.2.0.5.0，最后打补丁至 10.2.0.5.12。从版本角度讲，数据库就算安装完美了。

11g 和 9i/10g 不一样，如果原来的生产库是 11.2.0.3.0，可以升级至 11.2.0.4.0，再打补丁至 11.2.0.4.7。也可以另起一个 ORACLE_HOME，新安装一套 11.2.0.4.7 的软件，用这个新安装的软件，升级并启动数据库，最后可以删除老的 11.2.0.3 的软件。在全新安装的时候，不需要再安装 11.2.0.1 版本了，直接安装 11.2.0.4.0 版本，然后打一下 11.2.0.4.7 的补丁就可以了，前几个版本发布的介质都用不着了，而 9i/10g 必须要用 9.2.0.1/10.2.0.1 的介质。应该说 11g 在补丁方面比 9i/10g 做得要好。

本章节主要介绍升级小补丁 PSU。升级组件发行版本，我们放在数据库安装章节。

10.4.1 grid 打补丁 11.2.0.4.0 升级至 11.2.0.4.7

(1) 在 support.oracle.com 下载最新的补丁包，其中 patch:20760982 是数据库补丁集，patch:20996923 是 grid infrastructure 补丁集。还需要一个最新的 OPatch 11.2.0.3.11，也就是 p6880880_112000_Linux-x86-641.zip，按照 README 的要求，打 patch:20996923 这个补丁，OPatch 的版本至少是 OPatch utility version 11.2.0.3.6 或更高的版本，如图 10-1 所示。

复制至主机，这里放在 /oramed 下。

(2) 解压下载的补丁包。使用 root 用户，可以将下载的 3 个 zip 文件权限改为 777，对应的属性 grid 的补丁，权限改为 grid.oinstall，数据库的补丁权限改为 oracle.oinstall。

第10章 数据库的升级和补丁

11.2.0.4 Current Recommended Patches

Patch Set Updates

Document	Description	Rolling RAC	Patch Download
Note:21150851.8	Combo of 11.2.0.4.4 OJVM PSU and 11.2.0.4.7 DB PSU (Jul 2015)	Part	Patch:21150851
Note:21068539.8	Oracle JavaVM Component 11.2.0.4.4 Database PSU (Jul 2015) (OJVM PSU)	No	Patch:21068539
Note:20760982.8	11.2.0.4.7 (Jul 2015) Database Patch Set Update (DB PSU)	Yes	Patch:20760982
Note:19852360.8	Oracle JavaVM Component 11.2.0.4.1 Database PSU - Generic JDBC Patch (Oct 2014)	Yes	Patch:19852360

Grid Infrastructure

Document	Description	Rolling RAC	Patch Download
Note:21150864.8	Combo of 11.2.0.4.4 OJVM PSU and 11.2.0.4.7 GI PSU (Jul 2015)	Part	Patch:21150864
Note:20996923.8	11.2.0.4.7 (Jul 2015) Grid Infrastructure Patch Set Update (GI PSU)	Yes	Patch:20996923
Note:19852360.8	Oracle JavaVM Component 11.2.0.4.1 Database PSU - Generic JDBC Patch (Oct 2014)	Yes	Patch:19852360

图 10-1

```
cd /oramed/
[root@yingshuoramed]# chown -R grid.oinstall p20996923_112040_Linux-x86-64.zip
[root@yingshuoramed]# chown -R grid.oinstall p6880880_112000_Linux-x86-641.zip
[root@yingshuoramed]# chown -R oracle.oinstall p20760982_112040_Linux-x86-64.zip
[root@yingshuoramed]# chmod 777 p*.zip
[oracle@yingshuoramed]$ unzip p20996923_112040_Linux-x86-64.zip
[oracle@yingshuoramed]$ unzip p20760982_112040_Linux-x86-64.zip
[oracle@yingshuoramed]$ unzip p6880880_112000_Linux-x86-641.zip
```

(3) 更新 opatch。整个升级过程需要参考 readme 文档,readme 文档在补丁解压后的文件夹里。这里的 CRS_HOME 是/u01/app/grid/crs,要把新的 opatch 解压至这里面,将老的 opatch 废除。使用 root 用户先把 crs 这个文件夹的权限改为 777；然后将 opatch 压缩包复制至 CRS_HOME 下；将旧的 opatch 改名为 OPatch.bak；使用 grid 用户解压新的 opatch,新的 opatch 安装完毕。检查新的 opatch 版本。

```
[root@yingshu grid]# su - root
[root@yingshu ~]# cd /u01/app/grid/
[root@yingshu grid]# chmod 777 crs
su - grid
$ cp p6880880_112000_Linux-x86-641.zip $ORACLE_HOME/p6880880_112000_Linux-x86-641.zip
[grid@yingshucrs]$ mv OPatchOPatch.ba
[grid@yingshucrs]$ unzip p6880880_112000_Linux-x86-641.zip
[grid@yingshuOPatch]$ ./opatchlsinv
Oracle Interim Patch Installer version 11.2.0.3.11
Copyright (c) 2015, Oracle Corporation. All rights reserved.


Oracle Home       : /u01/app/grid/crs
Central Inventory : /u01/app/oraInventory
```

```
        from           : /u01/app/grid/crs//oraInst.loc
OPatch version         : 11.2.0.3.11
OUI version            : 11.2.0.4.0
```

(4) 将 grid 由 11.2.0.4.0 更新至 11.2.0.4.7。使用 oracle 用户停止 dbconsole：

```
su - oracle
emctl stop dbconsole
```

(5) 生成响应文件：

```
[grid@yingshuoramed] $ cd /oramed/20996923
[grid@yingshu 20996923] $ $ORACLE_HOME/OPatch/ocm/bin/emocmrsp - output /tmp/psu.rsp
OCM Installation Response Generator 10.3.7.0.0 - Production
Copyright (c) 2005, 2012, Oracle and/or its affiliates. All rights reserved.

Provide your email address to be informed of security issues, install and
initiate Oracle Configuration Manager. Easier for you if you use your My
Oracle Support Email address/User Name.
Visit http://www.oracle.com/support/policies.html for details.
Email address/User Name:

You have not provided an email address for notification of security issues.
Do you wish to remain uninformed of security issues ([Y]es, [N]o) [N]: y
The OCM configuration response file (/tmp/psu.rsp) was successfully created.
```

(6) 查看目前的软件安装情况，执行结果省略：

```
opatch lsinventory - detail - oh /u01/app/grid/crs/
```

(7) 使用 auto 进行安装升级，升级 rac 需要每个节点都执行一次 auto：

```
[root@yingshuinit.d]# su - grid
[grid@yingshu ~] $ su root
Password:
[root@yingshu grid]# opatch auto /oramed/20996923 - ocmrf /tmp/psu.rsp - oh /u01/app/grid/crs/
Executing /u01/app/grid/crs/perl/bin/perl /u01/app/grid/crs/OPatch/crs/patch11203.pl -
patchdir /oramed - patchn 20996923 - ocmrf /tmp/psu.rsp - oh /u01/app/grid/crs/ - paramfile /
u01/app/grid/crs/crs/install/crsconfig_params

This is the main log file: /u01/app/grid/crs/cfgtoollogs/opatchauto2015 - 07 - 29_11 - 46 - 45.log
This file will show your detected configuration and all the steps that opatchauto attempted to
do on your system:
/u01/app/grid/crs/cfgtoollogs/opatchauto2015 - 07 - 29_11 - 46 - 45.report.log

2015 - 07 - 29 11:46:45: Starting Oracle Restart Patch Setup
Using configuration parameter file: /u01/app/grid/crs/crs/install/crsconfig_params

Stopping CRS...
Stopped CRS successfully
```

```
patch /oramed/20996923/20760982 apply successful for home /u01/app/grid/crs
patch /oramed/20996923/20831122 apply successful for home /u01/app/grid/crs
patch /oramed/20996923/20299019 apply successful for home /u01/app/grid/crs
Starting CRS...
CRS-4123: Oracle High Availability Services has been started.
opatch auto succeeded.
```

(8) 验证升级结果：

```
[grid@yingshu ~]$ opatch lsinv
Oracle Interim Patch Installer version 11.2.0.3.11
Copyright (c) 2015, Oracle Corporation. All rights reserved.
... ...
Patch 20831122 : applied on Wed Jul 29 11:56:37 CST 2015
Patch description: "OCW Patch Set Update :11.2.0.4.7 (20831122)"
   Created on 1 Jul 2015, 06:26:45 hrs PST8PDT
   Bugs fixed:
     19270660, 18328800, 18691572, 20365005, 17750548, 17387214, 17617807
     14497275, 17733927, 18180541, 18962892, 17292250, 17378618, 16759171
... ...
Patch 20760982 : applied on Wed Jul 29 11:51:45 CST 2015
Unique Patch ID: 18908105
Patch description: "Database Patch Set Update :11.2.0.4.7 (20760982)"
   Created on 4 Jun 2015, 00:23:20 hrs PST8PDT
Sub-patch 20299013; "Database Patch Set Update : 11.2.0.4.6 (20299013)"
Sub-patch 19769489; "Database Patch Set Update : 11.2.0.4.5 (19769489)"
... ...
```

10.4.2 数据库打补丁 11.2.0.4.0 升级至 11.2.0.4.7

(1) ORACLE_HOME 下更新 opatch 请参见升级 grid 部分的"3 更新 opatch"章节。

(2) 整个升级过程需要参考 readme 文档,此文档在补丁解压后的文件夹里。

(3) 验证补丁冲突。

```
[oracle@yingshuoramed]$ cd20760982
[oracle@yingshu 20760982]$ opatchprereqCheckConflictAgainstOHWithDetail -ph ./
Oracle Interim Patch Installer version 11.2.0.3.10
Copyright (c) 2015, Oracle Corporation. All rights reserved.
Oracle Home : /oracle/app/oracle/product/10.2.0/db_1
... ...
Invoking prereq "checkconflictagainstohwithdetail"
Prereq "checkConflictAgainstOHWithDetail" passed.
OPatch succeeded.
```

(4) opatch apply 安装补丁,过程部分省略：

```
[oracle@yingshu ~]$ cd/oramed/20760982

[oracle@yingshu 20760982]$ opatch apply -local
OPatch continues with these patches: 17478514 18031668 18522509 19121551 19769489
20299013 20760982
```

```
Do you want to proceed? [y|n]
y
User Responded with: Y
All checks passed.
Provide your email address to be informed of security issues, install and
initiate Oracle Configuration Manager. Easier for you if you use your My
Oracle Support Email address/User Name.
Visit http://www.oracle.com/support/policies.html for details.
Email address/User Name:

You have not provided an email address for notification of security issues.
Do you wish to remain uninformed of security issues ([Y]es, [N]o) [N]: y

Please shutdown Oracle instances running out of this ORACLE_HOME on the local system.
(Oracle Home = '/oracle/app/oracle/product/10.2.0/db_1')

Is the local system ready for patching? [y|n]
y
… …
Composite patch 20760982 successfully applied.
Log file location: /oracle/app/oracle/product/10.2.0/db_1/cfgtoollogs/opatch/opatch2015-
07-29_13-28-33PM_1.log
OPatch succeeded.
```

(5) 执行 catbundle.sql psu apply：

```
[oracle@yingshu ~]$ sqlplus /nolog

SQL*Plus: Release 11.2.0.4.0 Production on Wed Jul 29 13:47:07 2015

Copyright (c) 1982, 2013, Oracle. All rights reserved.

SQL> conn /as sysdba
Connected to an idle instance.
SQL>
SQL> startup;
ORACLE instance started.

Total System Global Area 1023062016 bytes
Fixed Size                  2259520 bytes
Variable Size             318768576 bytes
Database Buffers          696254464 bytes
Redo Buffers                5779456 bytes
Database mounted.
Database opened.
SQL>@ $ORACLE_HOME/rdbms/admin/catbundle.sqlpsu apply
```

（6）重新编译失效对象：

```
sqlplus /nolog
conn /as sysdba
@?/rdbms/admin/utlrp.sql
```

（7）验证数据库补丁情况：

```
[oracle@yingshu ~]$ opatch lsinv
Oracle Interim Patch Installer version 11.2.0.3.10
Copyright (c) 2015, Oracle Corporation. All rights reserved.


Oracle Home       : /oracle/app/oracle/product/10.2.0/db_1
Interim patches (1) :

Patch  20760982     : applied on Wed Jul 29 13:35:01 CST 2015
Unique Patch ID: 18908105
Patch description: "Database Patch Set Update : 11.2.0.4.7 (20760982)"
    Created on 4 Jun 2015, 00:23:20 hrs PST8PDT
Sub-patch 20299013; "Database Patch Set Update : 11.2.0.4.6 (20299013)"
Sub-patch 19769489; "Database Patch Set Update : 11.2.0.4.5 (19769489)"
Sub-patch 19121551; "Database Patch Set Update : 11.2.0.4.4 (19121551)"
Sub-patch 18522509; "Database Patch Set Update : 11.2.0.4.3 (18522509)"
Sub-patch 18031668; "Database Patch Set Update : 11.2.0.4.2 (18031668)"
Sub-patch 17478514; "Database Patch Set Update : 11.2.0.4.1 (17478514)"
    Bugs fixed:
      17288409, 21051852, 18607546, 17205719, 17811429, 17816865, 20506699
      17922254, 17754782, 16934803, 13364795, 17311728, 17441661, 17284817
      16992075, 17446237, 14015842, 19972569, 17449815, 17375354, 19463897
      17982555, 17235750, 13866822, 18317531, 17478514, 18235390, 14338435
… …
--------------------------------------------------------------------------
OPatch succeeded.
```

（8）查看 DBA_REGISTRY_HISTORY 验证升级情况：

```
sqlplus /nolog
conn /as sysdba
set line 200
setpagesize 999
colaction_time for a30
col action for a10
col comments for a40
colobject_name for a25
colobject_type for a30
colcomp_name for a
colcomp_id for a10
col version for a12
col status for a8
selectcomp_id,comp_name,version,status from dba_registry;
```

```
COMP_ID        COMP_NAME                                VERSION      STATUS
----------     ------------------------------------     ----------   --------
OWM            Oracle Workspace Manager                 11.2.0.4.0   VALID
CATALOG        Oracle Database Catalog Views            11.2.0.4.0   VALID
CATPROC        Oracle Database Packages and Types       11.2.0.4.0   VALID

selectaction_time, action, comments from dba_registry_history;
SQL>
ACTION_TIME                      ACTION        COMMENTS
-------------------------        ----------    -------------------------
24-JUL-15 12.26.31.712545 PM     APPLY         Patchset 11.2.0.2.0
29-JUL-15 01.49.53.443522 PM     APPLY         PSU 11.2.0.4.7
29-JUL-15 01.50.55.103300 PM     APPLY         PSU 11.2.0.4.7
29-JUL-15 01.51.04.159587 PM     APPLY         PSU 11.2.0.4.7
```

10.5 版本升级总结

从力求完美的角度考虑,是要安装 PSU 补丁的,但有不少客户不安装 PSU 补丁,这无疑增加了出现故障的几率,问题可能就出现在察觉不到的地方,导致数据库运行变慢,响应迟钝。

升级过程中,主要的参考文档就是 readme。readme 中要求停止的,一定要停干净,停不干净可能导致升级失败,连原来的数据库软件都不能用了,结果有可能就是必须重装。

最好是在新装数据库软件时就安装补丁至最新,别等到投产之后再打补丁,毕竟动生产库和新装的数据库责任差别是巨大的。新安装时打补丁成本最低,生产库升级打补丁成本要高很多。

单机版的数据库升级比 RAC 要简单,RAC 需要先升级 grid 用户下的 CRS,再升级数据库。所以把数据库配成双机热备,对于数据库来说就是单机版,也是很好的选择,安装简单,维护成本低,出故障的几率小。

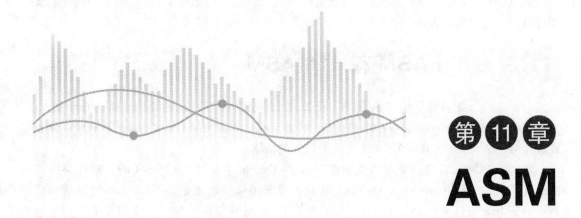

第 11 章 ASM

11.1 ASM 产生的背景

ASM 全称为 Automatic Storage Management(自动存储管理系统)。在海量数据库环境中,DBA 可能会花费很多的时间来做磁盘管理。例如一个表空间将占满整个磁盘,DBA 就需要再添加一块磁盘到操作系统中,然后再在新的磁盘上创建新的数据文件。如果是单个磁盘这倒不是很烦琐,问题是如果原先我们使用的是 RAID 或者是 LVM,那么现在大量的数据仍然是分布在以前的那些磁盘上,如果我们想让这些数据均匀地分布在以前的磁盘和新增加的磁盘上,可能就要耗费一天甚至几天的时间来做原先数据的导出导入。

11.2 ASM 的优势和特点

(1) 增加磁盘,减去旧磁盘变得非常简单。此特性可以实现轻松换存储。

(2) Oracle 直接垂直管理磁盘,省去了 VG、LV、文件系统等一切中间环节。

(3) 将文件分布到整个 diskgroup 中,实现最佳性能和最高的资源利用率,不浪费空间。

(4) 自动负载平衡功能,热盘和热块在 ASM 中很难出现,对整体 I/O 性能有质的提高。

(5) 统一 asmcmd、sqlplus、asmca、oem 工具管理,减少发生误操作。文件系统命令管不了 ASM。

(6) 条带化,以 files 的 allocation units (AUs)为单位的条带化。

(7) 可调整的均衡速度,可以使用 asm_power_limit 参数控制 ASM 中的数据的负载均衡速度,值为 0~11,数值越小均衡速度越慢,数值为 0 时表示不进行负载均衡,默认值为 1。

(8) 文件冗余,ASM 的镜像不需要第三方工具实现,提供三种级别的冗余方式。

(9) 自动的数据库文件管理使用 OMF(Oracle_Managed Files)方式管理文件(OMF 可以简化 DBA 的管理工作,不用指定文件的名字、大小、路径,由 Oracle 自动分配。在删除不再使用的文件时，OMF 也自动删除其对应的 OS 文件）。

（10）ASM 可以存储数据文件，控制文件，临时数据文件，数据文件 cope、spfile，日志文件，归档文件，回闪日志，RMAN 备份，expdp 导出的 dump 文件等。

11.3　10g ASM 和 11g ASM

10g ASM 是 ASM 的过渡版本，安装也比较麻烦，要安装 oracleasm-support…；oracleasm-…；oracleasmlib-…三个包，又不能放 voting disk 和 OCR 文件，还要使用裸设备放置这两个文件，所以 10g ASM 注定成为过渡的版本。

11g 第二版开始，我们就无法使用裸设备了，Oracle 不支持了。ASM 的稳定性易操作性也有了很大的长进，voting disk 和 OCR 盘再也不用单独建裸设备了。ASM 已经被集成到了第三个安装包里的 grid frastructure 里面了，不用再安装像 10g 一样的三个小程序包了，但我们要安装 grid frastructure 才能使用 ASM。如果是 RAC 必须安装 gird，如果是单机版的，就为一个 ASM 安装 grid，稍微有点浪费，不过我们也可以接受。11g 的 ASM 给大家的印象还是不错的。

11.4　ASM 双存储实验

我们先打开 asmca，单击 Create 按钮创建一个磁盘组，如图 11-1 所示。

图　11-1

这里我们假定 sdc 和 sdd 是一个存储，sde、sdf、sdg 是另外一个存储。我们实验的目的在于说明坏一个存储我们的 ASM 照样能正常运行。中间会用到 Failure Group 这个功能。

按图 11-2 的规则填写 diskgroup 内容。图 11-2 说明，第一个存储 sdc 和 sdd 为一个

failure group,名叫 failgrp1；第二个存储 sde 和 adf 为一个 failure group,名叫 failgrp2。单击 OK 按钮创建。

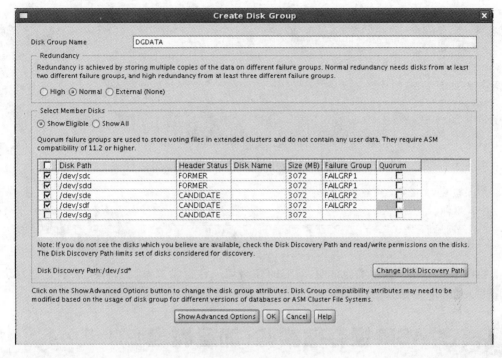

图 11-2

创建好之后,如图 11-3 所示。因为我们选择了 normal 冗余,我们的可用空间将损失一半。

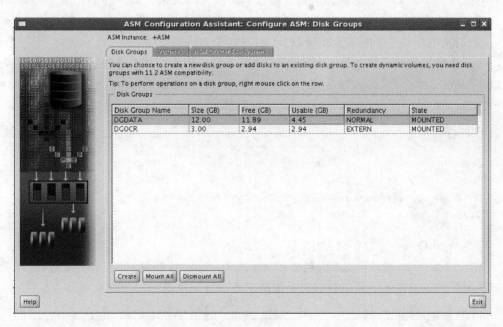

图 11-3

我们使用 dbca 在这个 DGDATA 磁盘组上创建数据库。具体过程略。

现在我们将第一个存储删除，模拟损坏，强制 dd 掉。

```
[root@yingshu ~]# dd if=/dev/zero of=/dev/sdc bs=8192 count=256
256+0 records in
256+0 records out
2097152 bytes (2.1 MB) copied, 0.001002 seconds, 2.1 GB/s
[root@yingshu ~]# dd if=/dev/zero of=/dev/sdd bs=8192 count=256
256+0 records in
256+0 records out
2097152 bytes (2.1 MB) copied, 0.000895 seconds, 2.3 GB/s
```

第一个存储已经坏了，磁盘全都不能用了，我们的数据库怎么样呢？

```
SQL> select open_mode from v$database;
OPEN_MODE
--------------------------------------------------------------
READ WRITE
```

数据库没事儿。这个实验证明，我们的存储坏一台也没关系。条件就是事先把冗余和 failure group 配置好。

11.5　ASM 换存储实验（加盘减盘）

ASM 可以轻松更换新旧存储，可谓易如反掌。如果想在文件系统或裸设备上换存储，那就麻烦了。下面就做一个简单的更换存储的实验。

先建一个 ASM 磁盘组，里面有两个磁盘 sdc 和 sdd，冗余方式为外部冗余 External，和 11.4 节中的存储规定一样，假定我们的存储一为旧存储。sde、sdf 和 sdg 为存储二（新存储）。我们先把存储二中的磁盘加到 DGDATA 这个磁盘组里去。

为了让磁盘间平衡数据的速度加快，在加盘减盘过程中有多个进程对其服务，我们先修改以下参数：

```
SQL> show parameter asm_power_limit
NAME                                 TYPE          VALUE
------------------------------------ ------------- ------------------------------
asm_power_limit                      integer       1
SQL>
SQL> alter system set asm_power_limit=11 scope=both sid='*';
System altered.
```

选择 Add Disks，添加存储，如图 11-4 所示。

如图 11-5 所示，选择第二个存储中的 sde、sdf、sdg 三块磁盘，单击 OK 按钮，这样我们的新存储就添加成功了。新旧存储的平衡情况查询以下视图。

```
select group_number,operation,state,power,actual,sofar,est_work,est_rate,est_minutes from v
  $asm_operation;
```

第11章 ASM

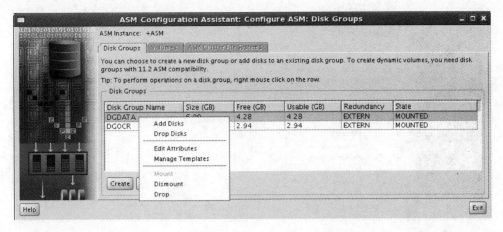

图 11-4

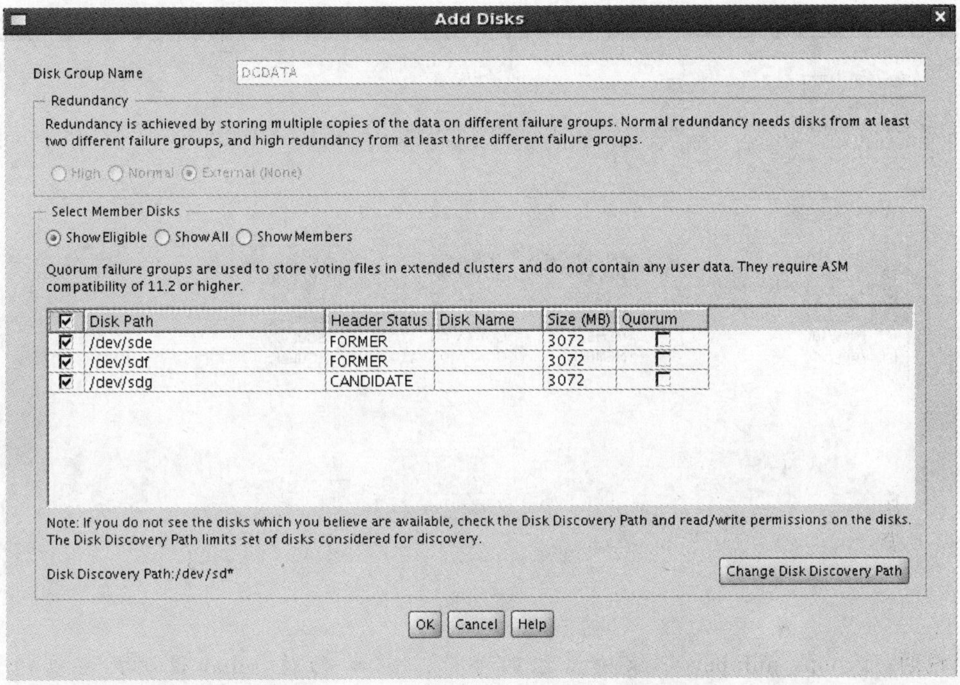

图 11-5

通过减盘的方式把旧存储在磁盘组里去掉。单击 Drop Disks,如图 11-6 所示。

如图 11-7 所示,选中旧存储中的磁盘 sdc、sdd,单击 OK 按钮删除。在实际的生产环境中,删除的过程可能比较慢。我们要关注以下视图,查看数据的移动情况。

```
SQL > select group_number,operation,state,power,actual,sofar,est_work,est_rate,est_minutes
from v$asm_operation;
GROUP_NUMBER OPERA STAT POWER ACTUAL SOFAR EST_WORK EST_RATE EST_MINUTES
------------ ----- ---- ----- ------ ----- -------- -------- -----------
           1 REBAL RUN     11     11    25      699      694           0
```

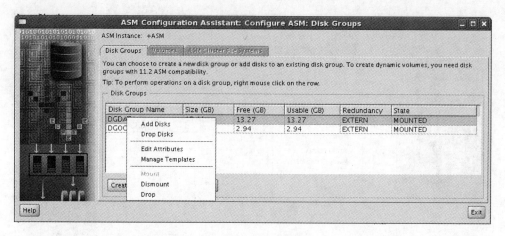

图 11-6

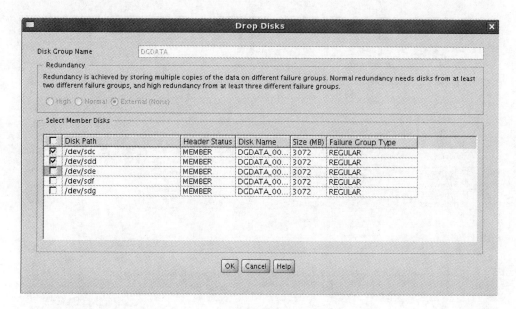

图 11-7

说明：group_number 记录磁盘组的序号；opera 的值 rebal 说明此磁盘组需要 rebalance；stat 说明 rebalance 正在进行；power 为负载均衡的速度；actual 为分配的速度；sofar 为已经移动了多少个 units；est_work 估算还要移动多少个 units；est_rate 估算每分钟移动多少个 units；est_minutes 估算还剩多少分钟 move 完成。

等这个视图 v$asm_operation 查不着数据了，我们的数据 move 工作才算完成了。这时候就可以把旧存储下电了。新旧存储的更换工作就算完成了。

11.6 与 ASM 相关的命令和视图

加盘减盘我们可以用 asmca，也可以用命令。命令能实现的功能远远大于图形界面，下面我们就介绍一些常用的命令。

用命令行的方式创建外部冗余的磁盘组：

```
SQL> conn /as sysasm
Connected.
SQL> create diskgroup dgidx external redundancy disk '/dev/sdd' name dginx_0001,'/dev/sde' name dgidx_0002;
Diskgroup created.
```

添加磁盘：

```
SQL> alter diskgroup dgidx add disk '/dev/sdf' name dgidx_0003;
Diskgroup altered.
```

删除磁盘：

```
SQL> alter diskgroup dgidx drop disk dgidx_0003;
Diskgroup altered.
```

删除磁盘组：

```
SQL> drop diskgroup dgidx;
Diskgroup created.
SQL> drop diskgroup dgidx including contents;
Diskgroup created.
```

创建标准冗余的磁盘组：

```
SQL> create diskgroup dgidx normal redundancy disk '/dev/sdd' name dgidx_v0001 ,'/dev/sde' name dgidx_v0002;
Diskgroup created.
```

装载磁盘组：

```
alter diskgroup dgdata mount;
Diskgroup altered.
```

卸载磁盘组：

```
SQL> alter diskgroup dgidx dismount;
Diskgroup altered.
```

手动的数据平衡在实际生产中很少用到，因为 ASM 会自动平衡，但我们可以使用 power 参数来控制平衡的速度。它的值为 0~11，可以根据 CPU 个数、磁盘的速度来确定参数的值：

```
SQL> conn /as sysasm
Connected.
SQL> alter diskgroup dgdata rebalance power 8;
Diskgroup altered.
```

查询当前的磁盘信息,包括可以使用但没使用的磁盘:

```
SQL> set linesize 132
SQL> col name for a20
sql> col failgroup for a20
sql> col path for a20
SQL> select group_number,disk_number,name,failgroup,create_date,path from v$asm_disk;

GROUP_NUMBER   DISK_NUMBER   NAME          FAILGROUP      CREATE_DA PATH
------------   -----------   -----------   -----------    --------- ---------
           0             0                                08-JUN-15 /dev/sdf
           0             2                                08-JUN-15 /dev/sdg
           2             0   DGOCR_0000    DGOCR_0000     27-MAY-15 /dev/sdb
           1             0   DGDATA_0000   DGDATA_0000
           3             0   DGIDX_V0001   DGIDX_V0001    08-JUN-15 /dev/sdd
           3             1   DGIDX_V0002   DGIDX_V0002    08-JUN-15 /dev/sde

6 rows selected.
```

查看当前的磁盘组信息:

```
SQL> select group_number,name,type,total_mb,free_mb from v$asm_diskgroup;

GROUP_NUMBER   NAME     TYPE      TOTAL_MB   FREE_MB
------------   ------   -------   --------   ---------
           2   DGOCR    EXTERN        3072        3013
           1   DGDATA   EXTERN        3072        1310
           3   DGIDX    NORMAL        6144        6042
```

查看当前 ASM 磁盘中的文件信息:

```
SQL> col TYPE for a20
SQL> select group_number,file_number,bytes,block_size,bytes/1024/1024 mb,type,creation_
date from v$asm_file;
GROUP_NUMBER   FILE_NUMBER   BYTES       BLOCK_SIZE   MB           TYPE             CREATION_
------------   -----------   ---------   ----------   ----------   --------------   ---------
           2           253        1536          512   .001464844   ASMPARAMETERFILE 27-MAY-15
           1           256     9748480        16384   9.296875     CONTROLFILE      08-JUN-15
           1           257    52429312          512   50.0004883   ONLINELOG        08-JUN-15
           1           258    52429312          512   50.0004883   ONLINELOG        08-JUN-15
           1           259    52429312          512   50.0004883   ONLINELOG        08-JUN-15
           1           260   734011392         8192   700.007813   DATAFILE         08-JUN-15
... ...
           1           265        2560          512   .002441406   PARAMETERFILE    08-JUN-15
11 rows selected.
```

查看 ASM 的别名信息,第三列代表是否是文件夹:

```
SQL> col name for a30
SQL> select group_number,file_number,alias_directory,name from v$asm_alias;
```

```
GROUP_NUMBER FILE_NUMBER A   NAME
------------ ----------- --- -------------------------------
           2  4294967295 Y   ASM
           2  4294967295 Y   ASMPARAMETERFILE
           2         253 N   REGISTRY.253.880830609
           1         256 N   Current.256.881852371
... ...
           1         265 N   spfileysdb.ora
20 rows selected.
```

校验磁盘组的一致性,若有错误就要修复:

```
alter diskgroup dgdata check all repair;
```

常用的 ASM 命令是基于 SQL*PLUS 管理 ASM。还可以通过 RAC 命令查看 ASM 的状态,如 crs_stat -t 与 crsctl stat res -t。通过 ASM 后台进程也可以判断 ASM 是否已经启动。

第12章 DataGuard

12.1 DataGuard 简介

DataGuard 是 Oracle 数据库最常用的容灾应急方式,它不像 GoldenGate 那样还需要单独购买软件许可,10g DataGuard 已包含在数据库企业版当中,不需要再购买 license。11g 是 active DataGuard,目标端可以查询。

如果说 RAC 的主要作用是防止主机故障,防止主机损坏;那么 DataGuard 的主要作用就是防止存储损坏。它可以轻易地、实时地将数据库复制到另外一个地方,可以同机房,如果网络带宽允许,也可以同城或者异地复制。

如果主库发生故障,备库可以实现快速的接管,接管的过程可以实现分钟级的接管,甚至更短的时间便可以实现接管主库。

DataGuard 可以实现一对多,一个主库,多个备库,最多可以有 9 个备库。一般情况下,有一个或两个备库也就够用了。也可以级联容灾,从 A 到 B,然后再从 B 到 C,配置方式灵活。

DataGuard 配置分类如下:

(1) 主库 primary,即生产库,单机或多节点 RAC 都可以。

(2) 备库 standby,即容灾库,单机或多节点都可以,但接收应用日志的节点只有一个。

(3) 逻辑 standby,将日志解析成 SQL 语句,然后在备库上执行这些 SQL 语句。逻辑 standby 有些数据类型不支持,如 bfile、encrypted columns、rowid、urowid、XMLType、对象类型、varrays、嵌套表、自定义类型;还有的 DML 语句不被支持或存储类型不被支持。优势就是在 9i、10g 的版本中,可以将数据库打开,作为查询库来使用。11g 版本中,物理 standby 也可以打开查询了,所以逻辑 standby 到了 11g 竞争力就小了。

(4) 物理 standby,将接收的归档日志恢复到备库上,备库和生产库完全一样,是主库的物理恢复,和 RMAN 备份恢复实质一样。是我们最常用的容灾模式。本书主要讨论物理

standby。

DataGuard 三种保护模式如下：

（1）最大保护模式 maximum protection。所有事务提交前不仅被写到本地的 online ~~redo log 中，同~~时还要提交到 standby 的 redo log 中，并确认其中一个 standby 可用，最后才会在 primary ~~中提~~交。

此种保护模~~式，因~~为备库的故障会影响主库的生产运营，所以几乎没人使用。

（2）最高性能 ~~maximum performa~~nce。事务可以随时提交，当前 primary 的 redo 也要至少写入一个 standby 数据库~~里，但不确认是否成功的~~。这是默认的保护模式。

（3）最高可用性 maximum availability。~~与最大保护模式的区别~~在~~于必须保障 redo 数据至少在一个 standby 数据库可用，不同与之不同的是，如果 无~~法写入 standby 数据库 redo log，primary 数据库并不会 shutdown，而是自动转为最高性能模式，待 standby 数据库恢复正常之后，它又会再自动转换成最高可用性模式。

```
SQL> select database_role,protection_mode,protection_level from v$database;

DATABASE_ROLE                    PROTECTION_MODE
----------------------------     ----------------------------
PROTECTION_LEVEL
----------------------------     ----------------------------
PRIMARY                          MAXIMUM PERFORMANCE
MAXIMUM PERFORMANCE
```

12.2　配置一个最常用的物理 DataGuard

在我们的样例里面，主库的名字叫 yingshudb1，备库的唯一名叫 yingshudg1。

12.2.1　将主库改为归档模式

```
[root@yingsu /]# su - oracle
[oracle@yingsu ~]$ ss

SQL*Plus: Release 11.2.0.4.0 Production on Wed Jan 14 08:59:08 2015

Copyright (c) 1982, 2013, Oracle. All rights reserved.


Connected to:
Oracle Database 11g Enterprise Edition Release 11.2.0.4.0 - 64bit Production
With the Partitioning, OLAP, Data Mining and Real Application Testing options

SQL>
SQL> alter system set log_archive_dest_1 = 'location=/arch1' scope=spfile;

System altered.
```

```
SQL>
SQL> shutdown immediate;
Database closed.
Database dismounted.
ORACLE instance shut down.
SQL> startup mount;
ORACLE instance started.

Total System Global Area  810106880 bytes
Fixed Size                  2257472 bytes
Variable Size             381685184 bytes
Database Buffers          419430400 bytes
Redo Buffers                6733824 bytes
Database mounted.
SQL> alter database archivelog;

Database altered.

SQL> alter database open;

Database altered.

SQL> alter system archive log current;

System altered.

SQL>
SQL> archive log list;
Database log mode              Archive Mode
Automatic archival             Enabled
Archive destination            /arch1
Oldest online log sequence     15
Next log sequence to archive   17
Current log sequence           17
```

注释：

```
[oracle@yingshu ~]$ ss
```

我们在环境变量里设置了别名，以后就省去了每次再去输入 sqlplus "/as sysdba"。

```
[oracle@yingshu ~]$ pwd
/home/oracle
[oracle@yingshu ~]$ cat .bash_profile
… …
export ORACLE_BASE=/u01/app/oracle/
export ORACLE_HOME=/u01/app/oracle/product/11.2.0/db_1
export ORACLE_SID=yingshudb1
alias ss="sqlplus '/as sysdba'"
```

经过上面的步骤，模式就设置好了。

12.2.2 将主库改为强制归档

```
SQL> alter database force logging;

Database altered.

SQL>
SQL> select force_logging from v$database;

FORCE_LOG
---------
YES
```

12.2.3 配置主库的 tnsnames.ora

```
[oracle@yingshu admin]$ cat tnsnames.ora
# tnsnames.ora Network Configuration File: /u01/app/oracle/product/11.2.0/db_1/network/admin/tnsnames.ora
# Generated by Oracle configuration tools.

YINGSDG1 =
  (DESCRIPTION =
    (ADDRESS_LIST =
      (ADDRESS = (PROTOCOL = TCP)(HOST = 192.168.1.20)(PORT = 1521))
    )
    (CONNECT_DATA =
      (SERVICE_NAME = yingsdg1)
    )
  )

YINGSDB1 =
  (DESCRIPTION =
    (ADDRESS_LIST =
      (ADDRESS = (PROTOCOL = TCP)(HOST = 192.168.1.10)(PORT = 1521))
    )
    (CONNECT_DATA =
      (SERVICE_NAME = yingsdb1)
    )
  )
```

可以用 netca 配置，也可以手工把连接选项写进 tnsnames.ora，我们要用 YINGSDG1 这个别名连接到远程数据库，将日志写到备库。

使用 tnsping 看看通不通。以下结果显示，我们的链路是通的。

```
[oracle@yingshu admin]$ tnsping yingsdg1
```

```
TNS Ping Utility for Linux: Version 11.2.0.4.0 - Production on 14-JAN-2015 11:45:16

Copyright (c) 1997, 2013, Oracle.  All rights reserved.

Used parameter files:
/u01/app/oracle/product/11.2.0/db_1/network/admin/sqlnet.ora


Used TNSNAMES adapter to resolve the alias
Attempting to contact (DESCRIPTION = (ADDRESS_LIST = (ADDRESS = (PROTOCOL = TCP)(HOST = 192.168.1.20)(PORT = 1521))) (CONNECT_DATA = (SERVICE_NAME = yingsdg1)))
OK (40 msec)
```

12.2.4 配置主库的参数

```
SQL> alter system set log_archive_config = 'dg_config=(yingsdb1,yingsdg1)' scope=both sid='*';

System altered.

SQL>
SQL> alter system set log_archive_dest_2 = 'service=yingsdg1 lgwr sync affirm valid_for=(online_logfiles,primary_role) db_unique_name=yingsdg1' scope=both sid='*';

System altered.

SQL> alter system set standby_file_management = auto scope=both sid='*';
System altered.

SQL> alter system set fal_server = 'yingsdg1' scope=both sid='*';
System altered.

SQL> alter system set fal_client = 'yingsdb1' scope=both sid='*';
System altered.

SQL> select * from v$dataguard_config;
DB_UNIQUE_NAME
------------------------------------------------------------------------
yingsdb1
yingsdg1

SQL> alter system set db_file_name_convert = '/u01/app/oracle/oradata/yingsdb1','/u01/app/oracle/oradata/yingsdb1' scope=spfile sid='*';

System altered.

SQL>
SQL> alter system set log_file_name_convert = '/u01/app/oracle/oradata/yingsdb1','/u01/app/oracle/oradata/yingsdb1' scope=spfile sid='*';

System altered.
```

第12章 DataGuard

注1：log_archive_dest_2='service=yingsdg1 lgwr sync affirm valid_for=(online_logfiles,primary_role) db_unique_name=yingsdg1' scope=both sid='*';

这个参数非常重要，它是我们 DataGuard 实现最重要的参数。

service 表示写到远程，如果是 location 一定是写到本地，如果是 service 一定是写到远程。

yingsdg1 是连接字符串的别名，tnsnames.ora 里面配置的别名，可以用 tnsping 测试是否通信正常。

lgwr 表示用什么写到远程，一般来说，推荐用 Log Writer Process（LGWR）来写远程，当然也可以用 archiver processes（ARCn）写到远程。也就是可以把 lgwr 改成 arch。

sync，是否同步写。如果改成 async 就是非同步写。非同步写可以降低对主库的性能的影响。

affirm 等待归档日志文件和 standby 日志文件同步完成之后，主库 log write 才能继续。noaffirm 主库的写进程，不等归档和备库日志写完便继续写主库日志。默认是 noaffirm。

valid_for=(online_logfiles,primary_role)。括号中，逗号前面是"写什么"，逗号后面是"在什么角色的时候写"。我们这里显示的就是，当此数据库为主库的时候，写在线日志文件。

db_unique_name=yingsdg1 这里要用数据库唯一名，区别主备库，因为物理 standby 主库和备库的数据库名是一样的，需要用数据库唯一名区别主备库。

注2：db_file_name_convert 参数，成对出现，只在此数据库为备库时起作用，逗号前面表示主库的数据文件所在的位置；逗号后面表示传到备库之后数据文件所放的位置。由此参数可以改变数据文件的位置。

12.2.5 备份主库、备份控制文件

注：如果 CPU 和存储允许，为加快备份速度，一定要多分几个通道。

```
[oracle@yingshu orabak]$ rman target /

Recovery Manager: Release 11.2.0.4.0 - Production on Mon Jan 19 10:06:15 2015

Copyright (c) 1982, 2011, Oracle and/or its affiliates. All rights reserved.

connected to target database: YINGSDB1 (DBID=4200332588)

RMAN> run {
2> allocate channel c1 type disk;
3> allocate channel c2 type disk;
4> allocate channel c3 type disk;
5> backup database
6> include current controlfile format '/orabak/orclfull_%d_%T_%s';
7> }

using target database control file instead of recovery catalog
allocated channel: c1
```

```
channel c1: SID = 1157 device type = DISK

allocated channel: c2
channel c2: SID = 24 device type = DISK

allocated channel: c3
channel c3: SID = 1159 device type = DISK

Starting backup at 19 - JAN - 15
channel c1: starting full datafile backup set
channel c1: specifying datafile(s) in backup set
... ...
piece handle = /orabak/orclfull_YINGSDB1_20150119_8 tag = TAG20150119T100656 comment = NONE
channel c2: backup set complete, elapsed time: 00:00:15
Finished backup at 19 - JAN - 15
released channel: c1
released channel: c2
released channel: c3
```

备份控制文件：控制文件一定是在全备之后备份，因为它里面要包含我们全备份时的信息。

```
RMAN > run {
2 > allocate channel c1 type disk;
3 > backup current controlfile for standby format '/orabak/control01.ctl.rman';
4 > }


allocated channel: c1
channel c1: SID = 1157 device type = DISK

Starting backup at 19 - JAN - 15
channel c1: starting full datafile backup set
channel c1: specifying datafile(s) in backup set
including standby control file in backup set
channel c1: starting piece 1 at 19 - JAN - 15
channel c1: finished piece 1 at 19 - JAN - 15
piece handle = /orabak/control01.ctl.rman tag = TAG20150119T100832 comment = NONE
channel c1: backup set complete, elapsed time: 00:00:01
Finished backup at 19 - JAN - 15
released channel: c1

RMAN >
```

12.2.6 拷贝所需的文件

将我们刚才的数据库全备份，备份的控制文件、参数文件、密码文件拷贝至备库。

```
[oracle@yingshu orabak]$ scp * oracle@192.168.1.20:/orabak/
Warning: Permanently added '192.168.1.20' (RSA) to the list of known hosts.
```

```
oracle@192.168.1.20's password:
control01.ctl.rman
… …
[oracle@yingshu orabak]$
```

建立对应的几个文件夹。使用 oracle 用户建立此文件夹。

```
mkdir -p /u01/app/oracle/diag/rdbms/yingsdg1/yingsdg1/trace
mkdir -p /u01/app/oracle/diag/rdbms/yingsdg1/yingsdg1/cdump
mkdir -p /u01/app/oracle/oradata/yingsdb1
mkdir -p /u01/app/oracle/admin/yingsdg1/adump
mkdir -p /arch1
```

12.2.7　启动备库的 listener

```
[oracle@yingshudg ~]$ lsnrctl
LSNRCTL > start
LSNRCTL > stat
Connecting to (DESCRIPTION = (ADDRESS = (PROTOCOL = TCP)(HOST = yingshudg)(PORT = 1521)))
STATUS of the LISTENER
------------------------
Alias                     LISTENER
Version                   TNSLSNR for Linux: Version 11.2.0.4.0 - Production
Start Date                19-JAN-2015 11:19:04
Uptime                    0 days 0 hr. 0 min. 6 sec
Trace Level               off
Security                  ON: Local OS Authentication
SNMP                      OFF
Listener Parameter File   /u01/app/oracle/product/11.2.0/db_1/network/admin/listener.ora
Listener Log File         /u01/app/oracle/diag/tnslsnr/yingshudg/listener/alert/log.xml
Listening Endpoints Summary...
  (DESCRIPTION = (ADDRESS = (PROTOCOL = tcp)(HOST = yingshudg)(PORT = 1521)))
  (DESCRIPTION = (ADDRESS = (PROTOCOL = ipc)(KEY = EXTPROC1521)))
Services Summary...
Service "yingsdb1" has 1 instance(s).
  Instance "yingsdg1", status READY, has 1 handler(s) for this service...
The command completed successfully
```

12.2.8　修改备库的 pfile

```
[oracle@yingshudg dbs]$ cat inityingsdg1.ora
yingsdb1.__oracle_base = '/u01/app/oracle/' # ORACLE_BASE set from environment
*.audit_file_dest = '/u01/app/oracle/admin/yingsdg1/adump'
*.audit_trail = 'db'
*.compatible = '11.2.0.4.0'
*.control_files = '/u01/app/oracle/oradata/yingsdb1/control01.ctl','/u01/app/oracle/oradata/yingsdb1/control02.ctl'
*.db_block_size = 8192
```

```
*.db_cache_size = 419430400
*.db_domain = ''
*.db_file_name_convert = '/u01/app/oracle/oradata/yingsdb1','/u01/app/oracle/oradata/yingsdb1'
*.db_name = 'yingsdb1'
*.diagnostic_dest = '/u01/app/oracle'
*.fal_client = 'yingsdb1'
*.fal_server = 'yingsdg1'
*.java_pool_size = 20971520
*.large_pool_size = 52428800
*.log_archive_dest_1 = 'location = /arch1'
*.log_file_name_convert = '/u01/app/oracle/oradata/yingsdb1','/u01/app/oracle/oradata/yingsdb1'
*.open_cursors = 300
*.pga_aggregate_target = 268435456
*.processes = 1500
*.remote_login_passwordfile = 'EXCLUSIVE'
*.shared_pool_size = 304087040
*.standby_file_management = 'AUTO'
*.undo_tablespace = 'UNDOTBS1'
######################
*.fal_server = 'yingsdb1'
*.fal_client = 'yingsdg1'
*.log_archive_dest_2 = 'service = yingsdb1 LGWR SYNC AFFIRM valid_for = (online_logfiles,primary_role) db_unique_name = yingsdb1'
*.log_archive_config = 'dg_config = (yingsdg1,yingsdb1)'
*.db_unique_name = 'yingsdg1'
*.service_names = 'yingsdg1'
```

注 1：####号以下部分，为配置 dg 重要参数，在备库的 pfile 中添加或修改正确。

注 2：其中 *.fal_server ＝ 'yingsdb1'配置的是对方的别名。

*.fal_client ＝ 'yingsdg1'中配置的是本机的别名。

这里配置的是数据库别名，也就是 tnsnames.ora 里面的连接字符串的别名。

注 3：初始化参数 LOG_ARCHIVE_CONFIG 用于控制发送归档日志到远程位置、接收远程归档日志。*.log_archive_config ＝ 'dg_config ＝ (yingsdg1,yingsdb1)'。这里配置的是数据库唯一名。

注 4：*.db_unique_name ＝ 'yingsdg1'这里的数据库唯一名是用来和主库做区分的。和主库一定不能一样。

注 5：*.db_name ＝ 'yingsdb1'备库的数据库名和主库的数据库名必须一样。

```
SQL> startup nomount ;
ORACLE instance started.

Total System Global Area  810106880 bytes
Fixed Size                  2257472 bytes
Variable Size             381685184 bytes
Database Buffers          419430400 bytes
Redo Buffers                6733824 bytes
```

12.2.9 恢复控制文件

```
RMAN > restore standby controlfile from '/orabak/control01.ctl.rman';
Starting restore at 19 - JAN - 15
using target database control file instead of recovery catalog
allocated channel: ORA_DISK_1
channel ORA_DISK_1: SID = 10 device type = DISK
channel ORA_DISK_1: restoring control file
channel ORA_DISK_1: restore complete, elapsed time: 00:00:01
output file name = /u01/app/oracle/oradata/yingsdb1/control01.ctl
output file name = /u01/app/oracle/oradata/yingsdb1/control02.ctl
Finished restore at 19 - JAN - 15
```

12.2.10 恢复备库

```
RMAN > alter database mount standby database;

database mounted
RMAN > run {
2 > allocate channel c1 type disk;
3 > allocate channel c2 type disk;
4 > allocate channel c3 type disk;
5 > restore database;
6 > }

allocated channel: c1
channel c1: SID = 13 device type = DISK

allocated channel: c2
channel c2: SID = 1146 device type = DISK

allocated channel: c3
channel c3: SID = 14 device type = DISK

Starting restore at 19 - JAN - 15

channel c1: starting datafile backup set restore
channel c1: specifying datafile(s) to restore from backup set
channel c1: restoring datafile 00002 to /u01/app/oracle/oradata/yingsdb1/sysaux01.dbf
channel c1: restoring datafile 00004 to /u01/app/oracle/oradata/yingsdb1/users01.dbf
channel c1: reading from backup piece /orabak/orclfull_YINGSDB1_20150119_6
channel c2: starting datafile backup set restore
channel c2: specifying datafile(s) to restore from backup set
channel c2: restoring datafile 00001 to /u01/app/oracle/oradata/yingsdb1/system01.dbf
channel c2: restoring datafile 00003 to /u01/app/oracle/oradata/yingsdb1/undotbs01.dbf
channel c2: reading from backup piece /orabak/orclfull_YINGSDB1_20150119_5

channel c1: piece handle = /orabak/orclfull_YINGSDB1_20150119_6 tag = TAG20150119T100656
```

```
channel c1: restored backup piece 1
channel c1: restore complete, elapsed time: 00:00:58
channel c2: piece handle = /orabak/orclfull_YINGSDB1_20150119_5 tag = TAG20150119T100656
channel c2: restored backup piece 1
channel c2: restore complete, elapsed time: 00:01:29
Finished restore at 19 - JAN - 15
released channel: c1
released channel: c2
released channel: c3
```

12.2.11 启动数据库

```
startup;
SQL > alter database recover managed standby database disconnect from session;

Database altered.
```

12.2.12 验证 DataGuard 两边是否同步

```
SQL > select max(sequence#) from v$archived_log;

MAX(SEQUENCE#)
--------------
            16

select max(sequence#) from v$log;

select max(sequence#) from v$log_history group by thread#;
```

12.2.13 DataGuard 相关的几个重要视图

```
select process,status,sequence# from v$managed_standby;
select sequence#, dest_id, archived, applied, deleted, status from v$archived_log order by
sequence#
```

12.3 主备切换 switch over

(1) 准备工作,首先查询主库状态,确认可以做切换:

```
SQL > select switchover_status from v$database;

SWITCHOVER_STATUS
------------------
TO STANDBY
```

(2) 在备库上面确认所有的 archive log 都已经 apply：

```
select sequence#,first_time,next_time,applied from v$archived_log;
```

(3) 在主库上 switch logfile：

```
Alter system switch logfile;或
alter system archive log current;
```

(4) 观察备库上已经 apply 所有 archive log：

```
select sequence#,first_time,next_time,applied from v$archived_log;
SEQUENCE#   FIRST_TIME      NEXT_TIME       APPLIED
---------   -----------    -----------    ---------
      160   28-FEB-15      28-FEB-15      YES
      161   28-FEB-15      28-FEB-15      YES
      162   28-FEB-15      28-FEB-15      YES
      163   28-FEB-15      28-FEB-15      YES
```

——切换：

(5) 将主库转变为可切换状态 primary 主库上执行：

```
alter database commit to switchover to physical standby with session shutdown;
```

(6) standby 备库：

```
alter database commit to switchover to primary;
```

(7) primary 主库：

```
shutdown immediate;
```

(8) standby 备库：

```
recover managed standby database cancel
shutdownimmediate;
startup
```

(9) 原来主库，新备库：

```
startup;
alter database recover managed standby database disconnect;
```

(10) 新的 primary：

```
startup;
alter system switch logfile;多执行几次.
```

查看 archive log 是否自动传输：
(11) 原来的 primary 新的 standby：

```
select sequence#,first_time,next_time,applied from v$archived_log;
select max(sequence#) from v$log;
```

12.4　FAILEOVER 切换实验

生产库一旦发生故障，我们可能就需要应急切换。或我们平时在日常演练中，也需要进行切换实验，以备紧急时刻我们的 datagurad 是完全可用的。下面我们就模拟生产库忽然损坏，而且不可恢复，由我们备库应急立刻顶住。整个切换过程也就几分钟的时间，把损失降到最低，把停机时间影响降到最低。

在模拟生产 down 机之前，我们先看看两边日志的同步情况。

（1）确认两边都是 215：

```
SQL> select max(sequence#) from v$log_history group by thread#;

MAX(SEQUENCE#)
--------------
           215
```

（2）停止 standby 数据库：

```
connect / as sysdba
SQL> alter database recover managed standby database finish;
Database altered.
```

（3）将 standby 数据库强制切换成主库：

```
SQL> alter database commit to switchover to primary;
Database altered.
```

（4）关闭 standby 数据库：

```
SQL> shutdown immediate
ORA-01109: database not open
Database dismounted.
ORACLE instance shut down.
```

（5）启动切换后的数据库：

```
SQL> startup
ORACLE instance started.

Total System Global Area 810106880 bytes
Fixed Size                  2257472 bytes
Variable Size             381685184 bytes
Database Buffers          419430400 bytes
Redo Buffers                6733824 bytes
Database mounted.
Database opened.
```

这时，我们的应急库就变成主库了，可以上生产了。

我们原来的主库修好之后，可以新配置成 DataGuard 备库。配置的过程参见 DataGuard 配置。

第 13 章 Oracle GoldenGate 实施参考

13.1 概述

Oracle GoldenGate 软件需要另外收费。GoldenGate 是创建于 1995 年的美国公司,开发总部设在旧金山,数据库复制领域的专业公司,2009 年被甲骨文 Oracle 公司收购,之后简称 ogg。和 DataGuard 一样,也是实现数据库容灾的一种技术。

GoldenGate 的优势在于:①可以实现异构的 IT 基础结构的容灾,不同数据库版本,不同操作系统,不同的数据库之间都可以。②对网络的压力和 DataGuard 相比要小。③对生产系统的影响较小。④可以实现数据库间的双向复制,但双向复制不能基于同一套表。⑤支持双活多活,复制的目标端也可以是 active 的。⑥过滤没必要的日志与事务。⑦数据迁移时可以实现零宕机。

DataGuard 的缺点是:①需要另行购买 licence。②对于大数据量的数据一致性效验存在疑问的,还需要购买 GoldenGate Veridata 来校验数据一致性。③对生产库有一些特殊要求,比如强烈推荐有主键或唯一索引,存在不支持的数据类型。④复制的实现比 DataGuard 要复杂。⑤对于 ddl 的复制暂只支持 Oracle to Oracle,况且对 ddl 的支持还需要另外的一些条件。⑥需要主机存储端配合的工作较多。比如存储队列的空间,如果是 RAC,要求将归档日志放在共享存储上。⑦需要保留一段时间的归档日志(建议为七天)。

13.2 深入了解 GoldenGate

数据容灾复制的技术很多,有存储级别的,很多银行金融机构在用;有数据库级别的 DataGuard、Quest Shareplex、Microsoft SQL Replication,还有 DSG(迪思杰)。Oracle 后来收购了 GoldenGate，所以摇身一变就成了 Oracle 官方支持的容灾软件了。

1．GoldenGate 的 5 个进程

（1）Capture（抓取进程）：实时读取 redo 变化，并可以实现过滤。

（2）Source trail（队列文件）：暂存数据的变化，可以设置保留天数。

（3）Pump（传输进程）：数据经过压缩和加密传送到目标端。

（4）DELIVERY（投递进程）：在目标数据库实现队列文件的应用提交。

（5）MANAGER（管理进程）：队列管理，进程管理，监控复制，接收 UI 的指令，提供延迟报告，事件报告，错误日志等，相当于总管家。

2．需要在磁盘上存储的组件

安装介质、Trail Files（队列文件）、Checkpoint Files（检查点文件）、Source Data Definition Files 数据定义文件、Configuration Files（配置文件）、Output Files（输出文件）、Report/Log Files（报告/日志文件）、Binaries 网络传送复本文件。

3．GoldenGate 的工作特点

（1）实时数据复制。

（2）支持异构环境。

（3）支持断点续传，不影响系统连续运行。

（4）高性能，对生产系统影响小。

（5）基于事件的复制，事务完整性，只复制提交的事务。

（6）功能强大的管理工具（ETL tools）、消息服务（Message Service）。

（7）灵活的拓扑结构。

（8）复制冲突检测和解决。

（9）事件标记基础架构（Event Marker Infrastructure）。

（10）路由和压缩：TCP/IP，LAN，WAN。

（11）传输过程加密：128-位，SSL。

（12）可以自定义延时应用修改，降低高峰时期的压力。

（13）根据事务大小和数量自动管理内存。

（14）分存式松散的成对关系。

（15）基于事务批量操作。

13.3 配置一个常用的 GoldenGate

1．系统现状调研

配置之前，我们要全面了解用户的现状。最好提前半个月甚至更长时间介入调研工作，研究适不适合用 ogg，如果决定使用 ogg，前期要做足准备工作，到真正实施的时候，就顺利多了。有的客户前期工作做的不充分，以至于在实施的过程中，困难重重。

▶ 我们需要了解的情况有：

（1）主库和备库的环境。

（2）操作系统列表。

第13章 Oracle GoldenGate实施参考

（3）数据库每天产生的日志量评估，峰值的日志量。

（4）网络带宽，如果带宽不足，是否启用压缩功能，以及压缩对 CPU 占用的影响。

（5）端口的开放情况。

（6）表结构，是否存在无主键表，是否存在不支持的数据类型。

（7）下载针对不同操作系统不同数据库版本的 ogg 软件。

（8）备库的数据库安装完毕。

（9）主库归档日志建议保留三天以上，如七天，以备追平日志。长事务需要最早记录的日志。

（10）主库如果是 RAC，归档日志必须放在共享存储上，共享空间要充足。

（11）ogg 队列日志建议保留至少三天，比如七天，以便出问题时续传。

（12）需要容灾复制的用户的列表，要复制的表的列表。

（13）确认是否复制 ddl，复制 ddl 可能会影响复制效率。

（14）主备库，CPU、内存大小和使用情况。

（15）主库需要另处建一个 ogg 表空间，建议预留空间，可以设为 10GB。

（16）无主键表或无唯一索引表建议建主键或唯一索引。

（17）确认同步过程所使用的技术，如 RMAN 或 expdp/impdp。

（18）测试同步 expdp/impdp 的时间，或了解 RMAN 备份恢复时间，由此估算实施时间。

2．配置 ogg 调查列表

表 13-1 是我们要进行配置 ogg 的调查列表。

表 13-1

	源　　端	目　标　端
操作系统版本	Redhat linux 6.1	Redhat linux 6.1
数据库版本	11.2.0.4	11.2.0.4
是否 RAC	NO	NO
IP 地址	192.168.1.10	192.168.1.80
服务器主机名	yingshu	ysogg
Oracle SID	ysdb	ysdbogg
是否 ASM	NO	NO
GG 版本	ogg112101	ogg112101
GG 安装所在用户	oracle	oracle
GG 安装所在主机	192.168.1.10	192.168.1.80
GG 安装路径	/goldengate/	/goldengate/
GoldenGate 数据库用户和表空间	goldengate	goldengate
复制的用户范围	共 1 个用户需要复制：	
特别说明	（1）关闭源端的 DDL 复制功能； （2）用户未加入复制的表； （3）用户下所有包括中文字段的表均不加入复制； （4）数据库发生的变更： ❶ 源端数据库的 recyclebin 关闭，以支持 DDL 复制； ❷ 源端数据库打开了数据库最小附加日志； ❸ 源端数据库对所有加入复制的表增加了表级附加日志	

3. 配置一个 GoldenGate

创建安装目录（主备库都要创建）：

```
[root@yingshu /]# mkdir /goldengate
[root@yingshu /]# chown -R oracle:dba /goldengate/
```

将下载完成的安装介质拷贝至/oramed/（主备库都要拷贝），解压安装介质（主备库都要解压）：

```
unzip ogg112101_fbo_ggs_Linux_x64_ora11g_64bit.zip -d /goldengate/
    cd /goldengate/
    tar -xvf fbo_ggs_Linux_x64_ora11g_64bit.tar
```

建立 ogg 表空间（主库创建）：

```
SQL> create tablespace goldengate datafile '/u01/app/oracle/oradata/ysdb/goldengate01.dbf' size 500m;
Tablespace created.
```

建立 ogg 用户（主库创建）：

```
create user ogg identified by ogg default tablespace goldengate;
User created.
grant resource,connect,dba to ogg;
Grant succeeded.
```

如果不想分给 ogg 用户 dba 权限，可以细化 ogg 的权限。分如下权限给 ogg：

```
grant connect to ogg;
grant alter any table to ogg;
grant alter session to ogg;
grant create session to ogg;
grant flashback any table to ogg;
grant select any dictionary to ogg;
grant select any table to ogg;
grant resource to ogg;
```

打开并检查主库附加日志（主库）：

```
alter database add supplemental log data ;
Database altered.
SQL> select
supplemental_log_data_min
,supplemental_log_data_pk
,supplemental_log_data_ui
,supplemental_log_data_fk
,supplemental_log_data_all from v$database;
```

SUPPLEMENTAL_LOG_DATA_MI	SUPPLEMEN	SUPPLEMEN	SUPPLEMEN	SUPPLEMEN
YES	NO	NO	NO	NO

第13章 Oracle GoldenGate实施参考

使用 oracle 用户,创建子目录(主备库都要创建):

```
[root@yingshu oramed]# su - oracle
[oracle@yingshu ~]$
[oracle@yingshu goldengate]$ ./ggsci

OracleGoldenGate Command Interpreter for Oracle
Version 11.2.1.0.1 OGGCORE_11.2.1.0.1_PLATFORMS_120423.0230_FBO
Linux, x64, 64bit (optimized), Oracle 11g on Apr 23 2012 08:32:14

Copyright (C) 1995, 2012, Oracle and/or its affiliates. All rights reserved.

GGSCI (yingshu) 1> create subdirs

Creating subdirectories under current directory /goldengate

Parameter files              /goldengate/dirprm: already exists
Report files                 /goldengate/dirrpt: created
Checkpoint files             /goldengate/dirchk: created
Process status files         /goldengate/dirpcs: created
SQL script files             /goldengate/dirsql: created
Database definitions files   /goldengate/dirdef: created
Extract data files           /goldengate/dirdat: created
Temporary files              /goldengate/dirtmp: created
Stdout files                 /goldengate/dirout: created
```

将数据库改成强制记日志 forceloging(主库):

```
SQL> alter database force logging;
Database altered.
SQL> selectforce_logging from v$database;
FORCE_LOG
---------
YES
```

关闭数据库的 recyclebin(仅实施 DDL 时进行配置)(主库):

```
SQL> alter session setrecyclebin = off;
Session altered.
show parameter recyclebin
recyclebin                          string           OFF
SQL> show parameterrecyclebin

NAME                       TYPE                 VALUE
-------------------------- -------------------- --------------------------
recyclebin                 string               OFF
```

添加环境变量(主备库都要添加):

```
export GG_HOME = /goldengate
export PATH = $PATH:$GG_HOME
export LIBPATH = $GG_HOME:$ORACLE_HOME/lib
```

编辑 GLOBALS 参数文件(主备库都要编辑):

```
edit params ./globals                    -- 在该文件中添加以下内容
ggschema ogg                             -- 指定的进行 ddl 复制的数据库用户
```

因为我们这里不复制 ddl,所以 ddl 的部分就省略了。

说明:我们这里假定生产用户有两个分别是 shu 和 ys; shu 用户有 stud 表,ys 用户有 tech 表。

```
SQL> connshu/shu
Connected.
SQL> selecttable_name from user_tables;
TABLE_NAME
--------------------------------------------------------------------------
STUD

SQL> connys/ys
Connected.
SQL> selecttable_name from user_tables;
TABLE_NAME
--------------------------------------------------------------------------
TEACH
```

打开表的补充日志(主库),所有需要复制的表,都需要打开:

```
GGSCI (yingshu) 1> dblogin userid ogg, password ogg
Successfully logged into database.
GGSCI (yingshu) 4> add trandata shu.stud
Logging of supplemental redo data enabled for table SHU.STUD.
GGSCI (yingshu) 5> add trandata ys.teach
Logging of supplemental redo data enabled for table YS.TEACH.
```

编辑 MGR 参数(源和目标均需要配置),为了简单,主备两边可以配成一样的。

```
[oracle@yingshu goldengate]$ ./ggsci
GGSCI (yingshu) 1> edit param mgr
PORT 7809
DYNAMICPORTLIST 7840 - 7914
PURGEOLDEXTRACTS ./dirdat/*, usecheckpoints, minkeepdays 1
LAGREPORTHOURS 1
LAGINFOMINUTES 30
LAGCRITICALMINUTES 45
PURGEDDLHISTORY MINKEEPDAYS 7, MAXKEEPDAYS 10
PURGEMARKERHISTORY MINKEEPDAYS 7, MAXKEEPDAYS 10
~
"dirprm/mgr.prm" [New] 8L, 247C written
```

启动管理进程 mgr 进程(源和目标均需要配置):

```
start mgr
```

查看管理进程 mgr 的状态：

```
GGSCI (yingshu) 12 > info all
Program     Status      Group       Lag atChkpt   Time Since Chkpt
MANAGER     RUNNING
```

添加源端的抽取 extract 进程（主库），begin now 表示从现在时刻开始抽取：

```
GGSCI (yingshu) 7 > add extract ggext, tranlog, begin now
EXTRACT added.
```

添加源端抽取进程生成的队列文件格式及位置（主库）：

```
GGSCI (yingshu) 1 > add exttrail ./dirdat/aa,extract ggext,megabytes 20
EXTTRAIL added.
```

后面的 megabytes 20，指抽取后生成队列文件的大小。

如果抽取进程添加错误，可以使用以下命令删除：

```
unregister extract ggext logretention
delete extract ggext !
```

添加抽取进程的内容，我们可以用 vi 在 ./dirprm 里直接编辑，也可以用 edit params ggext 在 ggsci 里编辑，两种方法的结果是一样的。编辑后的结果如下：

```
[oracle@yingshu dirprm] $ cat ggext.prm
EXTRACTggext
SETENV (NLS_LANG = AMERICAN_AMERICA.ZHS16GBK)
USERIDogg, PASSWORD ogg
EXTTRAIL ./dirdat/aa
TRANLOGOPTIONS ALTARCHIVELOGDEST PRIMARY INSTANCEysdb /arch1
GETTRUNCATES
table ys.*;
table shu.*;
```

启动抽取进程：

```
GGSCI (yingshu) 4 > start ggext
Sending START request toMANAGER ...
EXTRACT GGEXT starting
```

查看一下现在的状态，如果状态正常如下，我们就可以备份了。

```
GGSCI (yingshu) 6 >  info all
Program     Status      Group       Lag atChkpt   Time Since Chkpt
MANAGER     RUNNING
EXTRACT     RUNNING     GGEXT       00:00:00      00:00:05
```

我们还可以去 /goldengate/dirdat 下查看有没有生成以 aa 开头的队列文件，以验定我们添加的抽取进程是否正确。

我们先手工执行一下 crosscheck，实际生产环境中，根据实际需要执行。

```
crosscheck backup;
crosscheck archivelog all;
delete expired archivelog all;
report obsolete;
delete obsolete;
delete backup;
```

我们使用 RMAN 来备份,备份过程日志省略。备份脚本如下:

```
run{
backup database include current controlfile
format '/orabak/fullysdb_%d_%T_%s_%c';
sql 'alter system archive log current';
backup archivelog all
format '/orabak/ysdb_arch_%d_%T_%s_%c';
}
```

获取 SCN 号:

```
col status for a15
SQL> select group#, thread#, status, first_change#, next_change# from v$log;
    GROUP#    THREAD# STATUS      FIRST_CHANGE#    NEXT_CHANGE#
---------- ---------- ---------   -------------    -------------
         1          1 INACTIVE           267420           268989
         2          1 INACTIVE           268989           268997
         3          1 CURRENT            268997       2.8147E+14
```

说明:获取不活动的归档日志的最后一个 SCN 号,如果有多组 INACTIVE,取最大的 FIRST_CHANGE#号,本次取用的 SCN 号为 268989。

如果显示的 SCN 号为科学记数法,则用以下命令设置:

```
SQL> setnumw 15
```

使用 ftp 命令将备份的 db 和归档传到备库。我们这里在备库执行 mget,--后面为注释。

```
[root@ysogg /]# cd /orabak
[root@ysogg orabak]# ftp 192.168.1.10        -- 打开 ftp
ftp> cd /orabak                              -- 进入主库的/orabak 文件夹
ftp> prompt                                  -- 关闭提示
ftp> bin                                     -- 以二进制格式传输
ftp> mget *                                  -- 批量传输所有备份片
ftp> bye                                     -- 退出 ftp 工具
```

我们将主库的参数文件 spfile 和口令文件 orapwysdb 同样以 ftp 的方法,get 到备库的相应位置。

```
[oracle@ysogg ~]$ cd /u01/app/oracle/product/10.2.0/db_1/dbs
[oracle@ysogg dbs]$ ftp 192.168.1.10
ftp> cd /u01/app/oracle/product/10.2.0/db_1/dbs
```

```
ftp > bin
ftp > get orapwysdb
ftp > bye
```

还原数据文件（备库执行）：

```
RMAN > restore database;
```

在 RMAN 中恢复数据库（备库执行）：

```
run {
set until scn = 268989;
recover database;
}
```

以 regetlogs 方式打开数据库（备库执行）：

```
SQL > alter database open resetlogs;
Database altered.
```

我们可以将备库改成非归档模式（备库执行）：

```
Sql > shutdown immediate
Sql > start mount
Sql > alter database noarchivelog;
Sql > archive log list;
Sql > alter database open;
```

数据库正常打开，我们的数据同步就算完成了。

下面开始配置传输进程。

其实抽取进程也可以进行传输队列，但如果网络中断或数据库发生重启，抽取进程就要中断，重启后可能日志已经删除，无法正常抽取日志了。再有如果发生错误，不好区别是抽取的问题还是传输的问题。所以建议采用抽取和传输分开进程，分工明细，好定位好管理。

源端配 pump 进程（主库配置）：

```
GGSCI (yingshu) 54 > edit params ggpup      //也可以在/goldengate/dirprm 里编辑 ggpup.prm
                                            //文件
extract ggpup                               //定义抽取进程的名字
setenv (NLS_LANG = AMERICAN_AMERICA.ZHS16GBK) //设置字符集
passthru                                    //pass through 表示只传输不与库打交道
rmthost 192.168.1.80, mgrport 7809, compress //远程目标的地址端口号
rmttrail ./dirdat/aa                        //传到目标端后，放在什么位置
dynamicresolution                           //动态解析表结构，复制的表很多时，可以减少
                                            //启动时间
gettruncates                                //复制 truncate 表
table ys.tech;                              //用户名.表名，也可以用 * 号代替
table shu.stud;
```

在 ggsci 中添加传送进程(主库添加):

```
GGSCI (yingshu) 8 > add extract ggpup, exttrailsource ./dirdat/aa
EXTRACT added.
GGSCI (yingshu) 9 > add rmttrail ./dirdat/aa, extract ggpup, megabytes 20
RMTTRAIL added.
```

ggpup 是传输进程的名字,./dirdat/表示源队列存储的位置,aa 表示队列的名字。
如果添加错误或者想重新添加,可以把已添加的删除。

```
delete extract ggpup
```

创建检查点(备库创建):

```
edit params ./globals
checkpointtable .checktab



GGSCI (ysogg) 4 > dblogin userid ogg,password ogg
GGSCI (ysogg) 5 > add checkpointtable ogg.checktab
```

配置目标端 replicate 应用复写进程(备库配置)。

```
GGSCI (ysogg) 7 > add replicat ggrep, exttrail ./dirdat/aa
```

如果添加错误或需要重新配置,可以删除应用进程。

```
delete replicat ggrep
info replicat *,tasks
```

配置 replicate 复写进程的参数文件:

```
edit param ggrep
replicat   ggrep
setenv (NLS_LANG = AMERICAN_AMERICA.ZHS16GBK)
userid ogg, password ogg
handlecollisions                  //处理重复数据,必须有主键或唯一索引,多用于初始化
assumetargetdefs                  //两边数据结构,表结构一致用此参数
discardfile ./dirrpt/ggrep.dsc, append, megabytes 1000
gettruncates
map ys.* target ys.*;
map shu.* target shu.*;
```

启动应用进程:

```
start replicat ggrep
```

至此我们的 GoldenGate 就配置完成了。

第 14 章 常用Oracle工具在实际生产中的使用案例

14.1 10053 事件介绍及使用案例

1. 10053 事件介绍

10053 是 Oracle 官方非公开的一个事件,用来跟踪 Oracle 选择执行计划的过程,例如当发现一个 SQL 应该使用的索引,但却选择了全表扫描,或者没有选择正确的索引,可以采用 10053 事件来找到问题的根源所在。

10053 事件有以下两个级别:

(1) Level 2:2 级是 1 级的一个子集,它包含以下内容:

- ◆ Column statistics
- ◆ Single Access Paths
- ◆ Join Costs
- ◆ Table Joins Considered
- ◆ Join Methods Considered (NL/MS/HA)

(2) Level 1:1 级比 2 级更详细,它除了 2 级的内容外,还包括了:

- ◆ Parameters used by the optimizer
- ◆ Index statistics

启用 10053 事件的命令如下:

```
alter session set events = '10053 trace name context forever, level 1';
alter session set events = '10053 trace name context forever, level 2';
```

关闭 10053 事件：

```
alter session set events '10053 trace name context off';
```

2. 10053 诊断案例之索引选择不当导致 SQL 性能下降

（1）问题描述

在 AWR 报告里可见如下单次逻辑读较高的 SQL：

```
                       CPU     Elapsed
  Buffer Gets   Executions   per Exec  % Total   Time (s)   Time (s)   SQL Id
  ------------  ----------   --------  -------   --------   --------   --------
    71,412,938         652   109,529.0    27.3     639.27     671.98   2z49614gwr98z
SELECT A.CREDIT_NO, A.TRAN_TYPE, A.PAY_ACCNO, C.ACC_NAME, C.OPENACC_DEPT, A.REC
V_ACCNO, A.RECV_ACC_NAME, A.RECV_OPENACC_DEPT, A.AMOUNT, A.USEOF, A.CREDIT_NO, A
.STATUS, A.OPER_NAME, TO_CHAR(A.MAK_DATE, 'YYYY－MM－DD HH24:MI:SS'),A.CHECK_NAME,
TO_CHAR(A.CHK_DATE, 'YYYY－MM－DD HH24:MI:SS'),A.ADMIN_NAME, TO_CHAR(A.APP_DATE,

    24,292,778         230   105,620.8     9.3     217.71     218.02   2cnsd4n2tzdxq
SELECT A.CREDIT_NO, A.TRAN_TYPE, A.PAY_ACCNO, C.ACC_NAME, C.OPENACC_DEPT, A.REC
V_ACCNO, A.RECV_ACC_NAME, A.RECV_OPENACC_DEPT, A.AMOUNT, A.USEOF, A.CREDIT_NO, A
.STATUS, A.OPER_NAME, TO_CHAR(A.MAK_DATE, 'YYYY－MM－DD HH24:MI:SS'),A.CHECK_NAME,
TO_CHAR(A.CHK_DATE, 'YYYY－MM－DD HH24:MI:SS'),A.ADMIN_NAME, TO_CHAR(A.APP_DATE,

    14,604,128         141   103,575.4     5.6     128.52     128.92   g9823uvsvt121
SELECT A.CREDIT_NO, A.TRAN_TYPE, A.PAY_ACCNO, C.ACC_NAME, C.OPENACC_DEPT, A.REC
V_ACCNO, A.RECV_ACC_NAME, A.RECV_OPENACC_DEPT, A.AMOUNT, A.USEOF, A.CREDIT_NO, A
.STATUS, A.OPER_NAME, TO_CHAR(A.MAK_DATE, 'YYYY－MM－DD HH24:MI:SS'),A.CHECK_NAME,
TO_CHAR(A.CHK_DATE, 'YYYY－MM－DD HH24:MI:SS'),A.ADMIN_NAME, TO_CHAR(A.APP_DATE,

    11,524,114         107   107,702.0     4.4     103.30     123.25   ffsvcpfwfqcf1
SELECT A.CREDIT_NO, A.TRAN_TYPE, A.PAY_ACCNO, C.ACC_NAME, C.OPENACC_DEPT, A.REC
V_ACCNO, A.RECV_ACC_NAME, A.RECV_OPENACC_DEPT, A.AMOUNT, A.USEOF, A.CREDIT_NO, A
.STATUS, A.OPER_NAME, TO_CHAR(A.MAK_DATE, 'YYYY－MM－DD HH24:MI:SS'),A.CHECK_NAME,
TO_CHAR(A.CHK_DATE, 'YYYY－MM－DD HH24:MI:SS'),A.ADMIN_NAME, TO_CHAR(A.APP_DATE,
```

这是个典型的 OLTP 系统，4 个 SQL 的单次执行逻辑读都达到了 10 万以上，显然如此之高的逻辑读是不合理的，在执行次数不多的情况下，其影响不大，一旦执行次数大幅增加，必然带来严重的性能问题。

（2）问题分析

取一条完整的 SQL：

```
SELECT  A.CREDIT_NO, A.TRAN_TYPE, A.PAY_ACCNO, C.ACC_NAME, C.OPENACC_DEPT, A.RECV_ACCNO,
A.RECV_ACC_NAME, A.RECV_OPENACC_DEPT, A.AMOUNT, A.USEOF, A.CREDIT_NO, A.STATUS, A.OPER_NAME,
TO_CHAR(A.MAK_DATE, 'YYYY－MM－DD HH24:MI:SS'),A.CHECK_NAME,  TO_CHAR(A.CHK_DATE, 'YYYY－MM
－DDHH24:MI:SS'),
A.ADMIN_NAME, TO_CHAR(A.APP_DATE, 'YYYY－MM－DD HH24:MI:SS'),  A.MEMO1,   A.MEMO2,
A.FOLLOWINFO,  A.CANCEL_REASON, A.DEL_REASON,  A.CUR_TYPE,   0,TO_CHAR(A.ACC_DATE, 'YYYY－
MM－DDHH24:MI:SS'),
```

```
A.BALANCE,A.PAY_TYPE,A.TRANSFER_TYPE,A.FLAG,A.GENDAN FROM YINGSU_BARGAIN A, YINGSU_
ACCOUNT C
WHERE A.STATUS IN ('0','1','2','3','4','5','6','7','8','9','A','B','C') AND A.CUST_ID = RPAD(:a1,
21) AND
A.PAY_CUSTID = C.CUST_ID    AND A.TRAN_TYPE IN('0','1','2','3','A','R','P','S','D','I','O','G','F','W')
and A.PAY_ACCNO = C.ACC_NO    AND A.OPER_ID = RPAD(:a2,6) AND A.MAK_DATE between
to_DATE(:a3 || '000000','YYYYMMDDHH24MISS') and to_DATE(:a4 || '235959','YYYYMMDDHH24MISS')
AND A.PAY_ACCNO
= RPAD(:a5,32)    AND rownum < 5000    ORDER BY A.CREDIT_NO;
```

根据 mak_date 取值范围的不同，Oracle 优化器会产生两种不同的执行计划：

计划 1：

```
---------------------------------------------------------------------------------------
| Id  | Operation                      | Name                          | Rows | Bytes | Cost (%CPU)| Time     |
---------------------------------------------------------------------------------------
|  0  | SELECT STATEMENT               |                               |   1  |  470  |  10  (10)| 00:00:01 |
|* 1  |  FILTER                        |                               |      |       |          |          |
|  2  |   SORT ORDER BY                |                               |   1  |  470  |  10  (10)| 00:00:01 |
|* 3  |    COUNT STOPKEY               |                               |      |       |          |          |
|  4  |     NESTED LOOPS               |                               |   1  |  470  |   9   (0)| 00:00:01 |
|* 5  |      TABLE ACCESS BY INDEX ROWID| YINGSU_BARGAIN               |   1  |  368  |   7   (0)| 00:00:01 |
|* 6  |       INDEX RANGE SCAN         | YINGSU_BARGAIN_OPERID         |   4  |       |   3   (0)| 00:00:01 |
|  7  |      TABLE ACCESS BY INDEX ROWID| YINGSU_ACCOUNT               |   1  |  102  |   2   (0)| 00:00:01 |
|* 8  |       INDEX UNIQUE SCAN        | YINGSU_ACCOUNT_PK21026280981402|  1  |       |   1   (0)| 00:00:01 |
---------------------------------------------------------------------------------------
```

计划 2：

```
---------------------------------------------------------------------------------------
| Id  | Operation                      | Name                          | Rows | Bytes | Cost (%CPU)| Time     |
---------------------------------------------------------------------------------------
|  0  | SELECT STATEMENT               |                               |      |       |   9 (100)|          |
|  1  |  FILTER                        |                               |      |       |          |          |
|  2  |   SORT ORDER BY                |                               |   1  |  467  |   9  (12)| 00:00:01 |
|  3  |    COUNT STOPKEY               |                               |      |       |          |          |
|  4  |     NESTED LOOPS               |                               |   1  |  467  |   8   (0)| 00:00:01 |
|  5  |      TABLE ACCESS BY INDEX ROWID| YINGSU_BARGAIN               |   1  |  365  |   6   (0)| 00:00:01 |
|  6  |       INDEX RANGE SCAN         | YINGSU_BARGAIN_MAKEDATE       |   3  |       |   3   (0)| 00:00:01 |
|  7  |      TABLE ACCESS BY INDEX ROWID| YINGSU_ACCOUNT               |   1  |  102  |   2   (0)| 00:00:01 |
|  8  |       INDEX UNIQUE SCAN        | YINGSU_ACCOUNT_PK21026280981402|  1  |       |   1   (0)| 00:00:01 |
---------------------------------------------------------------------------------------
```

可以看出，这两种执行计划的不同之处在于选择了不同的索引，而优化器为什么会产生两种不同的选择，这取决于索引的选择率，而索引的选择率又是如何得到的呢？显然是 Oracle 优化器根据表及索引的统计信息计算得出的。考虑到数据库的统计信息是在每周六晚上进行收集，以 mak_date 这一列为例，在 20100227 晚间进行了数据库统计信息的收集，这样，统计信息里就记录了 mak_date 这一列的最大值和最小值，例如最大值为 20100227 22:00:00，最小值为 20020101 00:00:00。之后，当这条 SQL 再次被查询时，优化器会对这条 SQL 重新解析，产生新的执行计划，由于 mak_date 的值是采用绑定变量赋予的，在第一次解析这条 SQL 时，将对绑定变量值进行窥视，根据窥视到的值，产生一个合适的执行计划，这意味着第一次窥视到的绑定变量值决定了该 SQL 的执行计划。

在 mak_date 的查询条件中：

```
AND A.MAK_DATE between to_DATE(:a3 || '000000', 'YYYYMMDDHH24MISS') and to_DATE(:a4 || '235959',
'YYYYMMDDHH24MISS')
```

绑定变量:a3 和:a4 取值有多种可能，为了说明问题，可以简化为两种可能，一种是位于 20100227 22:00:00 和 20020101 00:00:00 这两个边界之间，另一种就是位于这两个边界之外，这两种情况，就产生了上述两种不同的执行计划。通过查看这两种情况下的 10053trace 文件，可以跟踪到优化器是如何根据不同的绑定变量窥视值来计算索引的选择率，进而根据索引的选择率，最终选择了走哪一个索引。

❶ 绑定变量值位于两个边界之间：

```
BIND#2
  oacdty = 96 mxl = 32(08) mxlc = 00 mal = 00 scl = 00 pre = 00
  oacflg = 03 fl2 = 1000000 frm = 01 csi = 01 siz = 0 off = 64
  kxsbbbfp = 9fffffffbf3de838   bln = 32   avl = 08   flg = 01
  value = "20100225"
Bind#3
  oacdty = 96 mxl = 32(08) mxlc = 00 mal = 00 scl = 00 pre = 00
  oacflg = 03 fl2 = 1000000 frm = 01 csi = 01 siz = 0 off = 96
  kxsbbbfp = 9fffffffbf3de858   bln = 32   avl = 08   flg = 01
  value = "20100226"
```

这种情况下，Oracle 优化器计算得到索引 YINGSU_BARGAIN_MAKEDATE 的选择率：

```
Access Path: index (RangeScan)
   Index: YINGSU_BARGAIN_MAKEDATE
   resc_io: 139.00   resc_cpu: 1225738
   ix_sel: 7.7741e-04   ix_sel_with_filters: 7.7741e-04
Cost: 139.46   Resp: 139.46   Degree: 1
```

经过与索引 yingsu_bargain_operid 对比，发现索引 yingsu_bargain_makedate 的成本较高，因而最后选择了走索引 yingsu_bargain_operid

```
   Access Path: index (AllEqRange)
   Index: YINGSU_BARGAIN_OPERID
   resc_io: 7.00   resc_cpu: 58198
   ix_sel: 2.3398e-05   ix_sel_with_filters: 2.3398e-05
   Cost: 7.02   Resp: 7.02   Degree: 1
   Best:: AccessPath: IndexRange   Index: YINGSU_BARGAIN_OPERID
         Cost: 7.02   Degree: 1   Resp: 7.02   Card: 0.00   Bytes: 0
```

❷ 绑定变量值位于边界之外：

```
Bind#2
  oacdty = 96 mxl = 32(08) mxlc = 00 mal = 00 scl = 00 pre = 00
  oacflg = 03 fl2 = 1000000 frm = 01 csi = 01 siz = 0 off = 64
  kxsbbbfp = 9fffffffbf3f1ea0   bln = 32   avl = 08   flg = 01
```

```
        value = "20100303"
Bind♯3
        oacdty = 96 mxl = 32(08) mxlc = 00 mal = 00 scl = 00 pre = 00
        oacflg = 03 fl2 = 1000000 frm = 01 csi = 01 siz = 0 off = 96
        kxsbbbfp = 9fffffffbf3f1ec0    bln = 32    avl = 08    flg = 01
        value = "20100304"
```

这种情况下,Oracle 优化器错误地估计了索引 yingsu_bargain_makedate 的选择率:

```
Using prorated density: 1.5749e-05 of col #3 as selectivity of out-of-range value pred
Access Path: index (RangeScan)
    Index: YINGSU_BARGAIN_MAKEDATE
    resc_io: 6.00    resc_cpu: 47796
    ix_sel: 1.5767e-05    ix_sel_with_filters: 1.5767e-05
Cost: 6.02    Resp: 6.02    Degree: 1
```

索引 yingsu_bargain_makedate 的选择率明显减小,而索引 yingsu_bargain_operid 的选择率没有发生变化:

```
Using prorated density: 7.3262e-05 of col #8 as selectivity of out-of-range value pred
Access Path: index (AllEqRange)
    Index: YINGSU_BARGAIN_OPERID
    resc_io: 7.00    resc_cpu: 58198
    ix_sel: 2.3398e-05    ix_sel_with_filters: 2.3398e-05
Cost: 7.02    Resp: 7.02    Degree: 1
```

最后,优化器选择了走索引 yingsu_bargain_makedate。

```
Best:: AccessPath: IndexRange    Index: YINGSU_BARGAIN_MAKEDATE
       Cost: 6.02    Degree: 1    Resp: 6.02    Card: 0.00    Bytes: 0
```

通过上述两种情况可以发现,Oracle 优化器会根据绑定变量窥视值以及统计信息提供的数据,来确定相应索引的选择率,对于在统计信息边界之内的绑定变量值,这种计算是接近事实的,而对于位于统计信息边界之外的绑定变量值,优化器给出了一个过低的 selectivity,而这个过低的 selectivity 是无法反映真实的数据分布情况的,这最终导致了 cost 计算错误,产生了不恰当的执行计划。

(3) 解决方案

在定位问题根源后,可以有以下解决方案:

❶ 对性能高的执行计划创建 SQL Profile,并绑定到 SQL 上,具体方法在后面有介绍。

❷ 删除索引 yingsu_bargain_makedate,这样优化器就选择了正确索引。

❸ 修改 SQL,加提示,强制使用效率较高的索引。

3. 10053 诊断案例之采样比例过低导致执行计划选择错误

某生产系统发现单个 SQL 耗时较高,占全库的 30.9%,AWR 里的报告如下:

```
    Elapsed       CPU                           Elap per        % Total
    Time (s)      Time (s)    Executions        Exec (s)        DB Time     SQL Id
    --------      --------    -----------       --------        --------    -------------
      11,089         504         31,166            0.4            30.9      acura616y0jnz
SELECT COUNT( * ) FROM YINGSU_DEL_TIME a,YINGSU_BARGAIN b WHERE a.credit_no = b
.credit_no AND b.cust_id = RPAD(:1,21) AND a.INFO_TYPE = '0' AND b.st
atus = 'A' AND b.cancel_reason is not null AND a.OPER_BROWSE = '0' AND
b.oper_id = RPAD(:2,6) union all SELECT COUNT( * ) FROM YINGSU_DEL_TIME a,BB_L
```

调查发现周六表统计信息收集后该 SQL 执行计划发生变化：

SNAP_ID	E_TIME	PLAN_HASH_VALUE	BUFFER_GETS_DELTA	DISK_READS_DELTA	EXECUTIONS_DELTA	OPTIMIZER_COST
23750	2010-10-23 23:00	3114794725	7	1	1	136
23751	2010-10-23 23:30	506460792	2330	2022	33	144

好的执行计划选择了 yingsu_bargain_his 表的 yingsu_bargain_his_cust_stat 索引：

```
Plan hash value: 3114794725

-----------------------------------------------------------------------------------------------------------
| Id | Operation                             | Name                         | Rows | Bytes | Cost (%CPU)| Time     | Pstart | Pstop |
-----------------------------------------------------------------------------------------------------------
|  0 | SELECT STATEMENT                      |                              |      |       |  57 (100) |          |        |       |
|  1 |  UNION-ALL                            |                              |      |       |           |          |        |       |
|  2 |   SORT AGGREGATE                      |                              |   1  |  62   |           |          |        |       |
|  3 |    NESTED LOOPS                       |                              |   1  |  62   |   9   (0) | 00:00:01 |        |       |
|  4 |     TABLE ACCESS BY INDEX ROWID       | YINGSU_BARGAIN               |   1  |  45   |   7   (0) | 00:00:01 |        |       |
|  5 |      INDEX RANGE SCAN                 | YINGSU_BARGAIN_STATUS        |   4  |       |   3   (0) | 00:00:01 |        |       |
|  6 |     TABLE ACCESS BY INDEX ROWID       | YINGSU_DEL_TIME              |   1  |  17   |   2   (0) | 00:00:01 |        |       |
|  7 |      INDEX UNIQUE SCAN                | IDX_YINGSU_DEL_TIME          |   1  |       |   1   (0) | 00:00:01 |        |       |
|  8 |   SORT AGGREGATE                      |                              |   1  |  63   |           |          |        |       |
|  9 |    NESTED LOOPS                       |                              |   1  |  63   |  48   (0) | 00:00:01 |        |       |
| 10 |     TABLE ACCESS BY GLOBAL INDEX ROWID| YINGSU_BARGAIN_HIS           |   1  |  46   |  46   (0) | 00:00:01 | ROW L  | ROW L |
| 11 |      INDEX RANGE SCAN                 | YINGSU_BARGAIN_HIS_CUST_STAT |  49  |       |   4   (0) | 00:00:01 |        |       |
| 12 |     TABLE ACCESS BY INDEX ROWID       | YINGSU_DEL_TIME              |   1  |  17   |   2   (0) | 00:00:01 |        |       |
| 13 |      INDEX UNIQUE SCAN                | IDX_YINGSU_DEL_TIME          |   1  |       |   1   (0) | 00:00:01 |        |       |
-----------------------------------------------------------------------------------------------------------
```

该执行计划运行时逻辑读和物理读都较低：

SNAP_ID	E_TIME	PLAN_HASH_VALUE	BUFFER_GETS_DELTA	DISK_READS_DELTA	EXECUTIONS_DELTA
23671	2014-10-22 07:30	3114794725	2513	660	254
23672	2014-10-22 08:00	3114794725	23101	6796	1374
23673	2014-10-22 08:30	3114794725	99472	29480	7131
23674	2014-10-22 09:00	3114794725	172454	46467	12888
23675	2014-10-22 09:30	3114794725	188126	55096	15037
23676	2014-10-22 10:01	3114794725	173177	51324	13805
23677	2014-10-22 10:30	3114794725	152507	46270	12515
23678	2014-10-22 11:00	3114794725	150677	47006	12501
23679	2014-10-22 11:30	3114794725	141880	40961	11454
23680	2014-10-22 12:00	3114794725	94010	27581	8142
23681	2014-10-22 12:30	3114794725	40787	11399	3854
23682	2014-10-22 13:00	3114794725	43900	11782	4102

差的执行计划选择了 yingsu_bargain_his 表的 yingsu_bargain_his_custmak 索引:

```
Plan hash value: 506460792

---------------------------------------------------------------------------------------------------
| Id | Operation                            | Name                     | Rows | Bytes | Cost (%CPU)| Time     | Pstart | Pstop |
---------------------------------------------------------------------------------------------------
|  0 | SELECT STATEMENT                     |                          |      |       | 176 (100)  |          |        |       |
|  1 |  UNION-ALL                           |                          |      |       |            |          |        |       |
|  2 |   SORT AGGREGATE                     |                          |    1 |    63 |            |          |        |       |
|  3 |    NESTED LOOPS                      |                          |    1 |    63 |   9   (0)  | 00:00:01 |        |       |
|  4 |     TABLE ACCESS BY INDEX ROWID      | YINGSU_BARGAIN           |    1 |    46 |   8   (0)  | 00:00:01 |        |       |
|  5 |      INDEX RANGE SCAN                | YINGSU_BARGAIN_STATUS    |    6 |       |   3   (0)  | 00:00:01 |        |       |
|  6 |     TABLE ACCESS BY INDEX ROWID      | YINGSU_DEL_TIME          |    1 |    17 |   1   (0)  | 00:00:01 |        |       |
|  7 |      INDEX UNIQUE SCAN               | IDX_YINGSU_DEL_TIME      |    1 |       |   0   (0)  |          |        |       |
|  8 |   SORT AGGREGATE                     |                          |    1 |    63 |            |          |        |       |
|  9 |    NESTED LOOPS                      |                          |    1 |    63 | 167   (1)  | 00:00:03 |        |       |
| 10 |     TABLE ACCESS BY GLOBAL INDEX ROWID| YINGSU_BARGAIN_HIS      |    1 |    46 | 166   (1)  | 00:00:02 | ROW L  | ROW L |
| 11 |      INDEX RANGE SCAN                | YINGSU_BARGAIN_HIS_CUSTMAK|  191 |       |   6   (0)  | 00:00:01 |        |       |
| 12 |     TABLE ACCESS BY INDEX ROWID      | YINGSU_DEL_TIME          |    1 |    17 |   1   (0)  | 00:00:01 |        |       |
| 13 |      INDEX UNIQUE SCAN               | IDX_YINGSU_DEL_TIME      |    1 |       |   0   (0)  |          |        |       |
---------------------------------------------------------------------------------------------------
```

该执行计划运行时逻辑读和物理读都较高:

SNAP_ID	E_TIME	PLAN_HASH_VALUE	BUFFER_GETS_DELTA	DISK_READS_DELTA	EXECUTIONS_DELTA
23809	2010-10-25 04:30	506460792	25	13	2
23810	2010-10-25 05:00	506460792	84	69	1
23811	2010-10-25 05:30	506460792	9	2	1
23812	2010-10-25 06:00	506460792	270	183	7
23813	2010-10-25 06:30	506460792	1103	927	12
23814	2010-10-25 07:01	506460792	5445	4443	54
23815	2010-10-25 07:30	506460792	56972	49209	191
23816	2010-10-25 08:00	506460792	411418	269546	1186
23817	2010-10-25 08:30	506460792	2848607	1075016	7033
23818	2010-10-25 09:00	506460792	5189883	2192857	14172
23819	2010-10-25 09:30	506460792	5196879	2703510	16994
23820	2010-10-25 10:00	506460792	4788809	2344351	15721
23821	2010-10-25 10:30	506460792	4008804	2247444	14569
23822	2010-10-25 11:00	506460792	3513755	1980171	13517
23823	2010-10-25 11:30	506460792	3326510	1930302	12831

索引情况如下:

```
sys@B2B > select index_name,column_position,column_name from dba_ind_columns where index_
name in ('YINGSU_BARGAIN_HIS_CUST_STAT','YINGSU_BARGAIN_HIS_CUSTMAK') and table_owner = 'B2C20'
order by 1,2;

INDEX_NAME                      COLUMN_POSITION      COLUMN_NAME
------------------------------  -------------------  ------------------------
YINGSU_BARGAIN_HIS_CUSTMAK                      1    CUST_ID
YINGSU_BARGAIN_HIS_CUSTMAK                      2    MAK_DATE
YINGSU_BARGAIN_HIS_CUST_STAT                    1    CUST_ID
YINGSU_BARGAIN_HIS_CUST_STAT                    2    STATUS
```

进一步分析索引相关统计信息,对于索引 YINGSU_BARGAIN_HIS_CUST_STAT,

发现 Oracle 收集的统计信息偏差较大，索引 YINGSU_BARGAIN_HIS_CUST_STAT 包含了两列 CUST_ID 和 STATUS，查询这两列的统计值：

```
sys@B2B> select table_name, column_name, num_distinct, density, num_nulls from dba_tab_
columns where  column_name in('CUST_ID','STATUS') and table_name = 'YINGSU_BARGAIN_HIS';

TABLE_NAME           COLUMN_NAME          NUM_DISTINCT     DENSITY       NUM_NULLS
-------------------  -------------------  ---------------  ------------  -----------
YINGSU_BARGAIN_HIS   CUST_ID              166607           6.0021E-06    0
YINGSU_BARGAIN_HIS   STATUS               5                .2            0
```

再查询该索引的统计值：

```
sys@B2B> select table_name, index_name, blevel, num_rows, leaf_blocks, distinct_keys, clustering_factor from dba_indexes where table_name = 'YINGSU_
BARGAIN_HIS' and index_name = 'YINGSU_BARGAIN_HIS_CUST_STAT';

TABLE_NAME           INDEX_NAME                    BLEVEL   NUM_ROWS  LEAF_BLOCKS  DISTINCT_KEYS  CLUSTERING_FACTOR
-------------------  ----------------------------  -------  --------  -----------  -------------  -----------------
YINGSU_BARGAIN_HIS   YINGSU_BARGAIN_HIS_CUST_STAT  3        23441010  193590       166607         19937240
```

以上是恢复旧的统计信息之后的情况，再看一下出问题时 10053 跟踪到的统计信息值：

```
Index: YINGSU_BARGAIN_HIS_CUST_STAT   Col#: 8 31
    LVLS: 3   #LB: 212050   #DK: 168439   LB/K: 3.00   DB/K: 340.00   CLUF: 21654270.00

Column (#8): CUST_ID(CHARACTER)
    AvgLen: 22.00 NDV: 168439 Nulls: 0 Density: 5.9369e-06
Column (#31): STATUS(CHARACTER)
    AvgLen: 2.00 NDV: 5 Nulls: 0 Density: 0.2
```

显然，无论是旧的统计信息，还是出问题的统计信息，索引的 distinct keys 值与 cust_id 的 NUM_DISTINCT 值完全一样，如果这是准确的，这意味着每个 cust_id 值只对应一种 status 值，从随机抽取的 cust_id 来看，一个 cust_id 都具有多个不同的 status 值：

```
sys@B2B> select distinct cust_id from b2c20.yingsu_bargain_his sample block(0.01);

CUST_ID
----------------------
P3300001008495#0L
P4100001000099#I
P4201000000138618#0
P4100001000313#1
P4201000000139864#K

12 rows selected.

sys@B2B> select distinct cust_id || status from b2c20.yingsu_bargain_his where cust_id = '
&cust_id'
  2 ;
Enter value for cust_id: P3300001008495#0L
old    1: select distinct cust_id || status from b2c20.yingsu_bargain_his where cust_id = '&cust_id'
```

```
New1: select distinct cust_id || status from b2c20.yingsu_bargain_his where cust_id =
'P3300001008495#0L'

CUST_ID || STATUS
-----------------------
P3300001008495#0L      5
P3300001008495#0L      A
P3300001008495#0L      4
P3300001008495#0L      2


sys@B2B>/
Enter value for cust_id: P4100001000099#I
old   1: select distinct cust_id || status from b2c20.yingsu_bargain_his where cust_id = '&cust_id'
new   1: select distinct cust_id || status from b2c20.yingsu_bargain_his where cust_id = '
P4100001000099#I'

CUST_ID || STATUS
-----------------------
P4100001000099#I       2
P4100001000099#I       4
P4100001000099#I       5
P4100001000099#I       A
```

显然，一个 cust_id 只具有一种 status 值的可能性是很小的，这意味着 Oracle 在收集该索引的 distinct keys 值时，存在着明显的缺陷，distinct keys 值被严重低估，从而直接导致优化器在选择索引时判断失误。

基于此，我们有理由怀疑这很可能是一个 bug，查询 metalink 可见与此相关的一个 bug，根据其描述，这是由于采用比例过低引起的。

Bug 9196440 Low distinct keys in index statistics (wrong scaling used) [ID 9196440.8]

14.2 10046 事件介绍及使用案例

1. 10046 事件介绍

SQL_TRACE 和 10046 是 Oracle 提供的用于进行 SQL 跟踪的工具，是日常的数据库故障诊断及调优的利器。关于如何开启 SQL_TRACE 及 10046 事件，这里就不做过多介绍，有兴趣的读者可以去网上搜索一下，有很多这方面的文章。下面主要介绍一个使用 10046 诊断事件的真实案例。

首先说明在 session 级开启 10046 的命令：

```
SQL> alter session set events '10046 trace name context forever, level 12';
Session altered.
```

关闭 10046 跟踪：

```
SQL> alter session set events '10046 trace name context off';
Session altered.
```

2. 使用 10046 进行 SQL 跟踪调优

生产上由于上了一个新的业务,出现了类似于这样的一条 SQL,取 t 表中查询满足 id 和 status 为某个值的所有记录中的第一条记录:

```
select id,status,comm from t where id = '1' and status = '2' and rownum = 1;
```

这个 SQL 的单次执行的逻辑读高达 1 万多,显然对于只取一条记录的 SQL 来说,这样的逻辑读是过高的。

检查这个表,发现只在 id 列上创建了一个索引,这样 Oracle 只能到表中对 status='2' 这个条件做 filter 操作,逻辑读大幅增加,而且,即使在这个 status 列的选择性不好的条件下,在这两列上创建复合索引的效率也远远大于在 id 列上的单列索引。为了说明此问题,可以构造一种比较极端的情况,进行测试。

构造一个表,共 50 万条记录,前 99999 条记录对应的 id 为 1,status 为 1,第 10 万条记录对应的 id 也为 1,但 status 为 2。这样,当以 id=1 and status=2 and rownum=1 为条件进行查询时,如果只在 id 上建立索引,那么 Oracle 将从索引中第一条 id=1 的记录开始扫描,每扫描一条索引记录就返回表中查找该记录中的 status 是否为 2,这样反复一直扫描到第 10 万条记录,才能找到第一条复合条件的记录并返回结果,这种情况下显然效率不高。而如果在 id 和 status 上面创建了复合索引,Oracle 只需要读取一条索引记录即可找到符合条件的记录,这种情况下查询效率与该索引的选择性实际上是没有关系的。

```
SQL> desc t
 Name                     Null?     Type
 ----                     ----      ----
 ID                                 CHAR(20)
 LEV                                VARCHAR2(10)
 BRANCH                             VARCHAR2(10)
 STATUS                             VARCHAR2(1)
 COMM                               VARCHAR2(100)
SQL> begin
  2    for i in 1 .. 500000 loop
  3      if i < 100000 then
  4        insert into t values('1',mod(100000,5),lpad(mod(i,20),5,'x'),'1','comm');
  5      elsif i = 100000 then
  6        insert into t values('1',mod(100000,5),lpad(mod(i,20),5,'x'),'2','comm');
  7      elsif i > 100000 and i <= 200000 then
  8        insert into t values('2',mod(100000,5),lpad(mod(i,20),5,'x'),'2','comm');
  9      elsif i > 200000 and i <= 300000 then
 10        insert into t values('3',mod(100000,5),lpad(mod(i,20),5,'x'),'2','comm');
 11      elsif i > 300000 and i <= 400000 then
 12        insert into t values('4',mod(100000,5),lpad(mod(i,20),5,'x'),'2','comm');
 13      else
 14        insert into t values('5',mod(100000,5),lpad(mod(i,20),5,'x'),'2','comm');
 15      end if;
```

```
 16    end loop;
 17   end;
 18   /
PL/SQL procedure successfully completed.
SQL> commit;
Commit complete.
SQL> select count(*) from t;

  COUNT(*)
----------
    500000

SQL> select count(*) from(select distinct id,status from t);
  COUNT(*)
----------
         6
```

```
SQL> select count(distinct id) from t;


COUNT(DISTINCTID)
-----------------
                5

SQL> create index idx_id on t(id) tablespace users;

Index created.

SQL> create index idx_id_status on t(id,status) tablespace users;

Index created.

SQL> select index_name,column_name,column_position from user_ind_columns where table_name = 'T';

INDEX_NAME              COLUMN_NAME              COLUMN_POSITION
----------------------  -----------------------  ---------------------
IDX_ID                  ID                                         1
IDX_ID_STATUS           ID                                         1
IDX_ID_STATUS           STATUS                                     2

SQL> select index_name,distinct_keys from user_indexes where table_name = 'T';

INDEX_NAME              DISTINCT_KEYS
--------------------------------------------
IDX_ID                              5
IDX_ID_STATUS                       6
```

当查询使用 id 列索引时：

```
SQL> select /*+ index(t idx_id) */ id,status,comm from t where id = '1' and status = '2' and
rownum = 1;

ID                    S COMM
--------------------  ---------
1                     2 comm

Execution Plan
----------------------------------------------------------
Plan hash value: 415658150

---------------------------------------------------------------------------------
| Id | Operation                    | Name   | Rows | Bytes | Cost (%CPU)| Time     |
---------------------------------------------------------------------------------
|  0 | SELECT STATEMENT             |        |    1 |    28 |     4   (0)| 00:00:01 |
|* 1 |  COUNT STOPKEY               |        |      |       |            |          |
|* 2 |   TABLE ACCESS BY INDEX ROWID| T      |    2 |    56 |     4   (0)| 00:00:01 |
|* 3 |    INDEX RANGE SCAN          | IDX_ID | 100K |       |     3   (0)| 00:00:01 |
---------------------------------------------------------------------------------

Predicate Information (identified by operation id):
---------------------------------------------------
   1 - filter(ROWNUM = 1)
   2 - filter("STATUS" = '2')
   3 - access("ID" = '1')

Statistics
----------------------------------------------------------
          0  recursive calls
          0  db block gets
        899  consistent gets
          0  physical reads
          0  redo size
        540  bytes sent via SQL*Net to client
        400  bytes received via SQL*Net from client
          2  SQL*Net roundtrips to/from client
          0  sorts (memory)
          0  sorts (disk)
          1  rows processed
```

逻辑读达到了 899，查看 10046 trace，可以清楚地看到 Oracle 一边读索引一边读表，obj♯=55900 为表，obj♯=55901 为索引：

```
……
WAIT #1: nam = 'db file sequential read' ela = 228 file# = 4 block# = 13227 blocks = 1 obj# =
55900 tim = 1356771811774771
WAIT #1: nam = 'db file sequential read' ela = 298 file# = 4 block# = 18829 blocks = 1 obj# =
55901 tim = 1356771811775194
WAIT #1: nam = 'db file sequential read' ela = 1095 file# = 4 block# = 13228 blocks = 1 obj# =
55900 tim = 1356771811776411
```

第14章 常用Oracle工具在实际生产中的使用案例

```
WAIT #1: nam = 'db file sequential read' ela = 123 file# = 4 block# = 18830 blocks = 1 obj# = 55901 tim = 1356771811776828
WAIT #1: nam = 'db file sequential read' ela = 535 file# = 4 block# = 13229 blocks = 1 obj# = 55900 tim = 1356771811777482
WAIT #1: nam = 'db file sequential read' ela = 115 file# = 4 block# = 13230 blocks = 1 obj# = 55900 tim = 1356771811777755
WAIT #1: nam = 'db file sequential read' ela = 380 file# = 4 block# = 18831 blocks = 1 obj# = 55901 tim = 1356771811778220
WAIT #1: nam = 'db file sequential read' ela = 943 file# = 4 block# = 13231 blocks = 1 obj# = 55900 tim = 1356771811779412
WAIT #1: nam = 'db file sequential read' ela = 607 file# = 4 block# = 18832 blocks = 1 obj# = 55901 tim = 1356771811780184
WAIT #1: nam = 'db file sequential read' ela = 556 file# = 4 block# = 13232 blocks = 1 obj# = 55900 tim = 1356771811781034
WAIT #1: nam = 'db file sequential read' ela = 779 file# = 4 block# = 18833 blocks = 1 obj# = 55901 tim = 1356771811782041
WAIT #1: nam = 'db file sequential read' ela = 1215 file# = 4 block# = 13233 blocks = 1 obj# = 55900 tim = 1356771811783368
WAIT #1: nam = 'db file sequential read' ela = 522 file# = 4 block# = 18834 blocks = 1 obj# = 55901 tim = 1356771811784071
WAIT #1: nam = 'db file sequential read' ela = 283 file# = 4 block# = 13234 blocks = 1 obj# = 55900 tim = 1356771811784434
WAIT #1: nam = 'db file sequential read' ela = 400 file# = 4 block# = 13235 blocks = 1 obj# = 55900 tim = 1356771811784991
WAIT #1: nam = 'db file sequential read' ela = 252 file# = 4 block# = 18835 blocks = 1 obj# = 55901 tim = 1356771811785346
WAIT #1: nam = 'db file sequential read' ela = 323 file# = 4 block# = 13236 blocks = 1 obj# = 55900 tim = 1356771811785840
WAIT #1: nam = 'db file sequential read' ela = 313 file# = 4 block# = 18836 blocks = 1 obj# = 55901 tim = 1356771811786263
WAIT #1: nam = 'db file sequential read' ela = 149 file# = 4 block# = 13237 blocks = 1 obj# = 55900 tim = 1356771811786528
WAIT #1: nam = 'db file sequential read' ela = 256 file# = 4 block# = 18837 blocks = 1 obj# = 55901 tim = 1356771811786913
WAIT #1: nam = 'db file sequential read' ela = 277 file# = 4 block# = 13238 blocks = 1 obj# = 55900 tim = 1356771811787278
WAIT #1: nam = 'db file sequential read' ela = 317 file# = 4 block# = 18838 blocks = 1 obj# = 55901 tim = 1356771811787778
WAIT #1: nam = 'db file sequential read' ela = 272 file# = 4 block# = 13239 blocks = 1 obj# = 55900 tim = 1356771811788153
… …
```

而当使用复合索引查询时，逻辑读明显降低：

```
SQL> select /* + index(t idx_id_status) */ id,status,comm from t where id = '1' and status = '2' and rownum = 1;

ID                   S COMM
---------------------
```

         1   2 comm

Execution Plan

Plan hash value: 3889392193

| Id | Operation | Name | Rows | Bytes | Cost (%CPU)| Time |

0	SELECT STATEMENT		1	28	4 (0)	00:00:01
* 1	COUNT STOPKEY					
2	TABLE ACCESS BY INDEX ROWID	T	2	56	4 (0)	00:00:01
* 3	INDEX RANGE SCAN	IDX_ID_STATUS	80360		3 (0)	00:00:01

Predicate Information (identified by operation id):
----

   1 - filter(ROWNUM = 1)
   3 - access("ID" = '1' AND "STATUS" = '2')

Statistics
----
          0 recursive calls
          0 db block gets
          4 consistent gets
          0 physical reads
          0 redo size
        540 bytes sent via SQL*Net to client
        400 bytes received via SQL*Net from client
          2 SQL*Net roundtrips to/from client
          0 sorts (memory)
          0 sorts (disk)
          1 rows processed

此时观察 10046 trace 文件，只有 4 次 db file sequential read 的等待产生，正好符合 4 个逻辑读的结果。

```
WAIT #1: nam = 'db file sequential read' ela = 277989 file# = 4 block# = 7772 blocks = 1 obj#
 = 55902 tim = 1356771851449041
WAIT #1: nam = 'db file sequential read' ela = 21429 file# = 4 block# = 21569 blocks = 1 obj#
 = 55902 tim = 1356771851471110
WAIT #1: nam = 'db file sequential read' ela = 6088 file# = 4 block# = 21365 blocks = 1 obj# =
 55902 tim = 1356771851478010
WAIT #1: nam = 'db file sequential read' ela = 16579 file# = 4 block# = 13628 blocks = 1 obj#
 = 55900 tim = 1356771851495606
```

通过对索引及表的数据块进行 dump 可以确认，其中前两行为索引 branch block，第三行为索引 leaf block，第四行为表的 data block。

第14章 常用Oracle工具在实际生产中的使用案例

```
SQL> select id,status,dbms_rowid.rowid_relative_fno(rowid),dbms_rowid.rowid_block_number
(rowid) from t where id = '1' and status = '2' and rownum = 1;

ID       S DBMS_ROWID.ROWID_RELATIVE_FNO(ROWID) DBMS_ROWID.ROWID_BLOCK_NUMBER(ROWID)
------   - ------------------------------------ ------------------------------------
1        2                                    4                                13628
```

符合条件的记录正好位于 file_id 为 4,block_id 为 13628 的数据块上。

通过上面的测试可以看出,对于这种 where 条件中有 rownum<=n 且 n 较小的查询,在 where 条件中所有的列上创建一个复合索引会明显地提高查询性能,其原因是因为只需要从索引中取满足条件的第一条记录,然后根据索引中的 rowid,回表中查询一次即可。

在这个测试实验中,通过 10046 跟踪事件,可以清晰地跟踪到 Oracle 在读取索引及表中数据块的顺序,从而可以更好地理解 Oracle 在执行 SQL 时是如何去读取相应的索引及表的数据块的。

14.3　SQL 优化利器之 SQL Profile 使用案例

1. SQL Profile 简介

SQL Profile 是 10g 的新特性,其本质上是一系列的 hints,SQL Profile 可以通过 Oracle 提供的 dbms_sqltune 来产生,也可以通过人为的方法去产生。当无法对应用 SQL 做修改时,可以通过使用 SQL Profile 来干预 SQL 的执行计划,从而在不改变 SQL 的情况下,使 SQL 的执行计划达到最优。

2. 使用 dbms_sqltune 来生成 SQL Profile 案例分析

生产库 CPU 使用率突然增加,发现一条 SQL 逻辑读较高,查看 SQL 的统计报告:

```
WORKLOAD REPOSITORY SQL Report
Snapshot Period Summary
DB Name        DB Id        Instance        Inst Num Release      RAC Host
------------ ---------- ------------ ----------------- ------------------
SMS          683801663  sms1                  1 10.2.0.5.0      YES mbsdb01
              Snap Id       Snap Time         Sessions    Curs/Sess
            -------------------------------------------------------
Begin Snap:   17692 18-Jun-14 15:00:19    764        19.0
  End Snap:   17693 18-Jun-14 15:30:33    760        19.1
   Elapsed:               30.23 (mins)
   DB Time:              166.57 (mins)
SQL Summary                         DB/Inst: SMS/sms1   Snaps: 17692-17693
                      Elapsed
    SQL Id          Time (ms)
----------- ----------
0c56xqjcayxv9     280,167
Module: JDBC Thin Client
SELECT A.UBANK_CODE,A.UBANK_NAME FROM YS_OTHER_BANK A,YS_AREA_DESC B WHERE SUBST
R(A.UBANK_CODE, 4, 4) = B.AREA_CODE AND B.AREA_CODE = :1 AND A.UBANK_CODE LIKE :
```

```
  2 ORDER BY A.UBANK_NAME

  -------------------------------------------------------------------

  SQL ID: 0c56xqjcayxv9            DB/Inst: SMS/sms1   Snaps: 17692-17693
  -> 1st Capture and Last Capture Snap IDs
     refer to Snapshot IDs witin the snapshot range
  -> SELECT A.UBANK_CODE,A.UBANK_NAME FROM YS_OTHER_BANK A,YS_AREA_DESC B W...

         Plan Hash        Total Elapsed                    1st Capture   Last Capture
  #      Value            Time(ms)         Executions      Snap ID       Snap ID
  --    -------------    --------------   -------------   -----------   --------------
  1      3843212254         280,167           3820           17693         17693

  -------------------------------------------------------------------


  Plan 1(PHV: 3843212254)
  -----------------------

  Plan Statistics                  DB/Inst: SMS/sms1   Snaps: 17692-17693
  -> % Total DB Time is the Elapsed Time of the SQL statement divided
     into the Total Database Time multiplied by 100

  Stat Name                          Statement        Per Execution       % Snap
  -------------------------------   -------------    ----------------   ----------
  Elapsed Time (ms)                   280,167             73.3              2.8
  CPU Time (ms)                       284,800             74.6              4.3
  Executions                            3,820             N/A               N/A
  Buffer Gets                         4.47E+07          11,697.0            31.8
  Disk Reads                               22              0.0              0.0
  Parse Calls                           3,804              1.0              0.0
  Rows                                237,731             62.2              N/A
  User I/O Wait Time (ms)                   5             N/A               N/A
  Cluster Wait Time (ms)                    2             N/A               N/A
  Application Wait Time (ms)                0             N/A               N/A
  Concurrency Wait Time (ms)              815             N/A               N/A
  Invalidations                             0             N/A               N/A
  Version Count                             2             N/A               N/A
  Sharable Mem(KB)                         23             N/A               N/A
  -------------------------------------------------------------------

  Execution Plan
  -------------------------------------------------------------------
  | Id | Operation                       | Name              | Rows | Bytes | Cost (%CPU)| Time     |
  -------------------------------------------------------------------
  |  0 | SELECT STATEMENT                |                   |      |       |   5 (100)  |          |
  |  1 |  SORT ORDER BY                  |                   |   1  |   53  |   5  (20)  | 00:00:01 |
  |  2 |   NESTED LOOPS                  |                   |   1  |   53  |   4   (0)  | 00:00:01 |
  |  3 |    TABLE ACCESS BY INDEX ROWID  | YS_OTHER_BANK     |   1  |   48  |   3   (0)  | 00:00:01 |
  |  4 |     INDEX RANGE SCAN            | YS_OTHER_BANK_PK01|   1  |       |   2   (0)  | 00:00:01 |
  |  5 |    INDEX RANGE SCAN             | YS_AREA_DESC_PK01 |   1  |    5  |   1   (0)  | 00:00:01 |
  -------------------------------------------------------------------
```

第14章 常用Oracle工具在实际生产中的使用案例

```
Full SQL Text

SQL ID        SQL Text
------------------------------------------------------------------------
0c56xqjcayxv  SELECT A.UBANK_CODE, A.UBANK_NAME FROM YS_OTHER_BANK A, BC_AREA_D
              DESC B WHERE SUBSTR(A.UBANK_CODE, 4, 4) = B.AREA_CODE AND B.AREA_
              CODE = :1 AND A.UBANK_CODE LIKE :2 ORDER BY A.UBANK_NAME


Report written to awrsqlrpt_1_17692_17693.txt
```

从 SQL 报告中看出,其单次执行的逻辑读达到 11 697,这在一个高并发性的 oltp 系统中是极其消耗资源的,首先怀疑表的统计信息是否有较大偏差:

```
SQL> select table_name,num_rows,to_char(last_analyzed,'yyyymmdd hh24:mi:ss')
  2    from user_tables
  3    where table_name in('YS_OTHER_BANK','YS_AREA_DESC');

TABLE_NAME                     NUM_ROWS   TO_CHAR(LAST_ANAL
------------------------------ ---------- -------------------
YS_AREA_DESC                         2077 20140614 23:00:19
YS_OTHER_BANK                      111488 20140614 23:03:15

sql> select count(*) from ys_area_desc;

  COUNT(*)
----------
      2077

sql> select count(*) from ys_other_bank;

  COUNT(*)
----------
    111117
```

检查发现统计信息是最新的,在这种情况下,重新收集统计信息之后,执行计划极有可能不会变化,因此,尝试使用 dbms_sqltune 来对 SQL 进行调优:

```
SQL> declare
  2    my_task_name varchar2(30);
  3  begin
  4    my_task_name := dbms_sqltune.create_tuning_task(
  5    begin_snap => 17692,
  6    end_snap => 17693,
  7    sql_id =>'0c56xqjcayxv9',
  8    plan_hash_value => 3843212254,
  9    scope =>'COMPREHENSIVE',
 10    time_limit => 60,
 11    task_name =>'0c56xqjcayxv9_20140619');
```

```
 12    end;
 13  /

PL/SQL procedure successfully completed.


SQL> begin
  2    dbms_sqltune.execute_tuning_task(task_name =>'0c56xqjcayxv9_20140619');
  3    end;
  4  /

PL/SQL procedure successfully completed.


SQL> set pages 1000
SQL> set long 10000
SQL> set longchunksize 10000
SQL> set lines 132
SQL> select dbms_sqltune.report_tuning_task('0c56xqjcayxv9_20140619') from dual;

DBMS_SQLTUNE.REPORT_TUNING_TASK('0C56XQJCAYXV9_20140619')
-------------------------------------------------------------------------------
GENERAL INFORMATION SECTION
-------------------------------------------------------------------------------
Tuning Task Name             : 0c56xqjcayxv9_20140619
Tuning Task Owner            : ORAMON
Scope                        : COMPREHENSIVE
Time Limit(seconds)          : 60
Completion Status            : COMPLETED
Started at                   : 06/19/2014 15:47:58
Completed at                 : 06/19/2014 15:48:00
Number of SQL Profile Findings : 1

-------------------------------------------------------------------------------
Schema Name: B2C20
SQL ID     : 0c56xqjcayxv9
SQL Text   : SELECT A.UBANK_CODE,A.UBANK_NAME FROM YS_OTHER_BANK
             A,YS_AREA_DESC B WHERE SUBSTR(A.UBANK_CODE, 4, 4) = B.AREA_CODE
             AND B.AREA_CODE = :1 AND A.UBANK_CODE LIKE :2 ORDER BY
             A.UBANK_NAME
-------------------------------------------------------------------------------
FINDINGS SECTION (1 finding)
-------------------------------------------------------------------------------

1SQL Profile Finding (see explain plans section below)
--------------------------------------------------------
  A potentially better execution plan was found for this statement.

  Recommendation (estimated benefit: 99.55%)
  ----------------------------------------
  Consider accepting the recommended SQLProfile.
    execute dbms_sqltune.accept_sql_profile(task_name =>
```

```
                '0c56xqjcayxv9_20140619', replace => TRUE);

-------------------------------------------------------------------
EXPLAIN PLANS SECTION
-------------------------------------------------------------------

1Original With Adjusted Cost
-----------------------------
Plan hash value: 3843212254

-------------------------------------------------------------------------------------
| Id | Operation                     | Name            | Rows  | Bytes | Cost (%CPU)| Time     |
-------------------------------------------------------------------------------------
|  0 | SELECT STATEMENT              |                 |    57 |  3021 | 14122   (1)| 00:02:50 |
|  1 |  SORT ORDER BY                |                 |    57 |  3021 | 14122   (1)| 00:02:50 |
|  2 |   NESTED LOOPS                |                 |    57 |  3021 | 14121   (1)| 00:02:50 |
|  3 |    TABLE ACCESS BY INDEX ROWID| DYS_OTHER_BANK  | 13282 |  622K |   832   (1)| 00:00:10 |
|* 4 |     INDEX RANGE SCAN          | YS_OTHER_BANK_PK01 | 13132 |    |    59   (0)| 00:00:01 |
|* 5 |    INDEX RANGE SCAN           | YS_AREA_DESC_PK01|   1 |     5 |     1   (0)| 00:00:01 |
-------------------------------------------------------------------------------------

Predicate Information (identified by operation id):
---------------------------------------------------

   4 access("A"."UBANK_CODE" LIKE :2)
       filter("A"."UBANK_CODE" LIKE :2)
   5 access("B"."AREA_CODE" = :1)
       filter("B"."AREA_CODE" = SUBSTR("A"."UBANK_CODE",4,4))

2Using SQL Profile
-------------------
Plan hash value: 2722777387

-------------------------------------------------------------------------------------
| Id | Operation                     | Name            | Rows  | Bytes | Cost (%CPU)| Time     |
-------------------------------------------------------------------------------------
|  0 | SELECT STATEMENT              |                 |    57 |  3021 |    62   (2)| 00:00:01 |
|  1 |  SORT ORDER BY                |                 |    57 |  3021 |    62   (2)| 00:00:01 |
|  2 |   TABLE ACCESS BY INDEX ROWID | YS_OTHER_BANK   |     6 |   288 |    59   (0)| 00:00:01 |
|  3 |    NESTED LOOPS               |                 |    57 |  3021 |    61   (0)| 00:00:01 |
|* 4 |     INDEX RANGE SCAN          | YS_AREA_DESC_PK01|   1 |     5 |     2   (0)| 00:00:01 |
|* 5 |     INDEX RANGE SCAN          | YS_OTHER_BANK_PK01|   6 |      |    58   (0)| 00:00:01 |
-------------------------------------------------------------------------------------

Predicate Information (identified by operation id):
---------------------------------------------------

   4 access("B"."AREA_CODE" = :1)
   5 access("A"."UBANK_CODE" LIKE :2)
       filter("A"."UBANK_CODE" LIKE :2 AND "B"."AREA_CODE" = SUBSTR("A"."UBANK_CODE",4,4))
```

可以看出，Oracle 生成了一个 SQL Profile，并给出了使用它的具体语句：

```
execute dbms_sqltune.accept_sql_profile(task_name =>'0c56xqjcayxv9_20140619', replace =>
TRUE);
```

在使用此 SQL Profile 后，SQL 的逻辑读由 10000 多降低到了 100 以内，服务器 CPU 使用率明显下降。

3．手动生成 SQL Profile

在某些情况下，使用 dbms_sqltune 无法产生 SQL Profile，此时如果我们可以通过手动产生一个 SQL Profile，并将其绑定到有问题的 SQL 上，下面是一个示范。

首先看一个查询：

```
sys@YSBANK> select object_name from t where object_id between 1000 and 2000;
1001 rows selected.

Execution Plan
----------------------------------------------------------
Plan hash value: 1601196873
----------------------------------------------------------
| Id  | Operation          | Name | Rows  | Bytes | Cost (%CPU)| Time     |
----------------------------------------------------------
|   0 | SELECT STATEMENT   |      | 15384 |  345K |   533   (1)| 00:00:07 |
|*  1 |  TABLE ACCESS FULL | T    | 15384 |  345K |   533   (1)| 00:00:07 |
----------------------------------------------------------

Predicate Information (identified by operation id):
----------------------------------------------------------
   1 - filter("OBJECT_ID"<=2000 AND "OBJECT_ID">=1000)

Statistics
----------------------------------------------------------
          1  recursive calls
          0  db block gets
       2387  consistent gets
          0  physical reads
      29996  redo size
      27644  bytes sent via SQL*Net to client
       1107  bytes received via SQL*Net from client
         68  SQL*Net roundtrips to/from client
          0  sorts (memory)
          0  sorts (disk)
       1001  rows processed
```

对于这个查询，由于表中统计信息过旧，错误地选择了全表扫描，加索引提示：

```
sys@YSBANK>
sys@YSBANK> select /*+ index(t idx_t) */ object_name from t where object_id between 1000
and 2000;
1001 rows selected.
```

```
Execution Plan
----------------------------------------------------------
Plan hash value: 1594971208
----------------------------------------------------------
| Id  | Operation                   | Name  | Rows  | Bytes | Cost (%CPU)| Time     |
----------------------------------------------------------
|   0 | SELECT STATEMENT            |       | 15384 |  345K | 15421   (1)| 00:03:06 |
|   1 |  TABLE ACCESS BY INDEX ROWID| T     | 15384 |  345K | 15421   (1)| 00:03:06 |
| * 2 |   INDEX RANGE SCAN          | IDX_T | 15384 |       |    35   (0)| 00:00:01 |
----------------------------------------------------------

Predicate Information (identified by operation id):
----------------------------------------------------------
   2 access("OBJECT_ID">=1000 AND "OBJECT_ID"<=2000)


Statistics
----------------------------------------------------------
          0  recursive calls
          0  db block gets
        203  consistent gets
          0  physical reads
          0  redo size
      27644  bytes sent via SQL*Net to client
       1107  bytes received via SQL*Net from client
         68  SQL*Net roundtrips to/from client
          0  sorts (memory)
          0  sorts (disk)
       1001  rows processed
```

对于上述查询,显然加索引提示是合理的,而当无法修改应用时,我们需要从数据库端来干预此 SQL 的执行计划,采用 SQL Profile 可以完成此功能。

我们暂且把未加提示的 SQL 称之为 sql_nohi,加了提示的 SQL 称之为 sql_hi,执行下面的脚本 gen_sql_profile.sql,该脚本的功能是首先取得 sql_hi 的 outline 数据,也就是一系列 hints,然后使用这些 outline 数据来对 sql_nohi 生成 profile,也就是说让 sql_nohi 来使用 sql_hi 的执行计划:

```
declare
  ar_hint_table     sys.dbms_debug_vc2coll;
  ar_profile_hints sys.sqlprof_attr := sys.sqlprof_attr();
  cl_sql_text       clob;
  i                 pls_integer;
begin
  with a as (
  select
          rownum as r_no
        , a.*
  from
          table(
            -- replace with
```

```
                     -- DBMS_XPLAN.DISPLAY_AWR
                     -- if required
                     dbms_xplan.display_cursor(
                       '&&good_sql_id'
                     , null
                     , 'OUTLINE'
                     )
                     -- dbms_xplan.display_awr(
                     --     '&&1'
                     -- , null
                     -- , null
                     -- , 'OUTLINE'
                     -- )
                 ) a
    ),
    b as (
    select
            min(r_no) as start_r_no
    from
            a
    where
            a.plan_table_output = 'Outline Data'
    ),
    c as (
    select
            min(r_no) as end_r_no
    from
            a
          , b
    where
            a.r_no > b.start_r_no
    and     a.plan_table_output = '    */'
    ),
    d as (
    select
            instr(a.plan_table_output, 'BEGIN_OUTLINE_DATA') as start_col
    from
            a
          , b
    where
            r_no = b.start_r_no + 4
    )
    select
            substr(a.plan_table_output, d.start_col) as outline_hints
    bulk collect
    into
            ar_hint_table
    from
            a
          , b
```

```
              , c
              , d
    where
              a.r_no >= b.start_r_no + 4
    and       a.r_no <= c.end_r_no 1
    order by
              a.r_no;

    select
              sql_text
    into
              cl_sql_text
    from
              -- replace with dba_hist_sqltext
              -- if required for AWR based
              -- execution
              v$sql
              -- sys.dba_hist_sqltext
    where
              sql_id = '&&bad_sql_id';

-- this is only required
-- to concatenate hints
-- splitted across several lines
-- and could be done in SQL, too
i := ar_hint_table.first;
while i is not null
loop
  if ar_hint_table.exists(i + 1) then
    if substr(ar_hint_table(i + 1), 1, 1) = ' ' then
      ar_hint_table(i) := ar_hint_table(i) || trim(ar_hint_table(i + 1));
      ar_hint_table.delete(i + 1);
    end if;
  end if;
  i := ar_hint_table.next(i);
end loop;

i := ar_hint_table.first;
while i is not null
loop
  ar_profile_hints.extend;
  ar_profile_hints(ar_profile_hints.count) := ar_hint_table(i);
  i := ar_hint_table.next(i);
end loop;

dbms_sqltune.import_sql_profile(
   sql_text     => cl_sql_text
 , profile      => ar_profile_hints
 , name         => 'PROFILE_&&1'
 -- use force_match => true
```

```
        -- to use CURSOR_SHARING = SIMILAR
        -- behaviour, i.e. match even with
        -- differing literals
     , force_match => false
     );
end;
/
```

此过程可以完成上述功能,测试如下:

首先取得两条 SQL 的 sql_id:

```
sys@YSBANK> select sql_id,sql_text from v$sql where sql_text like ('select % from t where
object_id between 1000 and 2000 %');
SQL_ID          SQL_TEXT
--------------- --------------------------------------------------------------
dmn81fk77n9z6   select object_name from t where object_id between 1000 and 2
                000
036w3z13k3wgh   select /*+ index(t idx_t) */ object_name from t where object
                _id between 1000 and 2000
```

执行 gen_sql_profile.sql:

```
sys@YSBANK>@gen_sql_profile.sql
Enter value for good_sql_id: 036w3z13k3wgh
old   17:                '&&good_sql_id'
new   17:                '036w3z13k3wgh'
Enter value for bad_sql_id: dmn81fk77n9z6
old   83:           sql_id = '&&bad_sql_id';
new   83:           sql_id = 'dmn81fk77n9z6';
Enter value for profile_name: test
old  112:     , name        => 'PROFILE_&profile_name'
new  112:     , name        => 'PROFILE_test'
PL/SQL procedure successfully completed.
```

此时再执行未加 hints 的 sql:

```
sys@YSBANK> set autot traceonly
sys@YSBANK> select object_name from t where object_id between 1000 and 2000;
1001 rows selected.

Execution Plan
----------------------------------------------------------
Plan hash value: 1594971208
--------------------------------------------------------------------------------
| Id  | Operation                    | Name  | Rows  | Bytes | Cost (%CPU)| Time     |
--------------------------------------------------------------------------------
|   0 | SELECT STATEMENT             |       | 15384 |  345K | 15421   (1)| 00:03:06 |
|   1 |  TABLE ACCESS BY INDEX ROWID | T     | 15384 |  345K | 15421   (1)| 00:03:06 |
|*  2 |   INDEX RANGE SCAN           | IDX_T | 15384 |       |    35   (0)| 00:00:01 |
--------------------------------------------------------------------------------
```

```
Predicate Information (identified by operation id):
---------------------------------------------------
   2 access("OBJECT_ID">=1000 AND "OBJECT_ID"<=2000)
Note
-----
   SQL profile "PROFILE_test" used for this statement

Statistics
----------------------------------------------------------
          1  recursive calls
          0  db block gets
        203  consistent gets
          0  physical reads
          0  redo size
      27644  bytes sent via SQL*Net to client
       1106  bytes received via SQL*Net from client
         68  SQL*Net roundtrips to/from client
          0  sorts (memory)
          0  sorts (disk)
       1001  rows processed
```

可以看到，Oracle 使用了 SQL Profile，SQL 的执行计划得以改变，对于上述 SQL Profile，是 SQL_id 相同的条件下生效的，改变一下 where 条件中的值：

```
sys@YSBANK> exec dbms_sqltune.drop_sql_profile('PROFILE_test');
PL/SQL procedure successfully completed.
sys@YSBANK>@gen_sql_profile
Enter value for good_sql_id: 036w3z13k3wgh
old   17:              '&&good_sql_id'
new   17:              '036w3z13k3wgh'
Enter value for bad_sql_id: dmn81fk77n9z6
old   83:          sql_id = '&&bad_sql_id';
new   83:          sql_id = 'dmn81fk77n9z6';
Enter value for profile_name: TEST
old  112:    , name        => 'PROFILE_&profile_name'
new  112:    , name        => 'PROFILE_TEST'
PL/SQL procedure successfully completed.
sys@YSBANK> set autot traceonly explain
sys@YSBANK> select object_name from t where object_id between 1000 and 2001;
Execution Plan
----------------------------------------------------------
Plan hash value: 1594971208

--------------------------------------------------------------------------------
| Id | Operation                   | Name  | Rows  | Bytes | Cost (%CPU)| Time     |
--------------------------------------------------------------------------------
|  0 | SELECT STATEMENT            |       | 15399 |  345K | 15436   (1)| 00:03:06|
|  1 |  TABLE ACCESS BY INDEX ROWID| T     | 15399 |  345K | 15436   (1)| 00:03:06|
|* 2 |   INDEX RANGE SCAN          | IDX_T | 15399 |       |    35   (0)| 00:00:01|
--------------------------------------------------------------------------------
```

```
Predicate Information (identified by operation id):
---------------------------------------------------
   2 access("OBJECT_ID">=1000 AND "OBJECT_ID"<=2001)
Note
-----
   SQL profile "PROFILE_TEST" used for this statement
sys@YSBANK>select object_name from t where object_id between 1000 and 3000;
Execution Plan
----------------------------------------------------------
Plan hash value: 1594971208
----------------------------------------------------------------------------
| Id | Operation                     | Name  | Rows  | Bytes | Cost (%CPU)| Time     |
----------------------------------------------------------------------------
|  0 | SELECT STATEMENT              |       | 30736 |  690K | 30809   (1)| 00:06:10 |
|  1 |  TABLE ACCESS BY INDEX ROWID  | T     | 30736 |  690K | 30809   (1)| 00:06:10 |
|* 2 |   INDEX RANGE SCAN            | IDX_T | 30736 |       |    69   (0)| 00:00:01 |
----------------------------------------------------------------------------
Predicate Information (identified by operation id):
---------------------------------------------------
   2 access("OBJECT_ID">=1000 AND "OBJECT_ID"<=3000)
Note
-----
   SQL profile "PROFILE_TEST" used for this statement
```

可见,对于这种没有使用绑定变量的 SQL,即使是 where 条件中的取值不同,仍然能够使用 SQL Profile,这是由于设置了 force_match => true 的原因。

第 15 章

Oracle 12c 介绍

15.1 Oracle Database 12c 简介

2013 年 6 月 26 日，Oracle Database 12c 版本正式发布，Oracle 数据库 12c 引入了一个新的多承租方架构，使用该架构可轻松部署和管理数据库云。此外，一些创新特性可最大限度地提高资源使用率和灵活性，如 Oracle Multitenant 可快速整合多个数据库，而 Automatic Data Optimization 和 Heat Map 能以更高的密度压缩数据和对数据分层。这些独一无二的技术进步再加上在可用性、安全性和大数据支持方面的主要增强，使得 Oracle 数据库 12c 成为私有云和公有云部署的理想平台。

Oracle 12c 增加了 500 多项新功能，其新特性主要涵盖了 6 个方面：云端数据库整合的全新多租户架构、数据自动优化、深度安全防护、面向数据库云的最大可用性、高效的数据库管理以及简化大数据分析。可插播数据库允许我们将数百个数据库整合到一个能够确保独立性的 RAC 环境，而之前我们必须将其放在单独的服务器上。毫无疑问，Oracle Database 12c 最吸引人的特性就是支持整合。可插拔数据库可以共享内存资源，由于这些数据库仍旧是单个数据库实例，因此能够简化管理。

15.2 Oracle 12c 体系结构

图 15-1 为 Oracle 12c Pluggable Database 的体系结构示意图，在 Oracle 12c 中引入了 CDB 与 PDB 的新特性，在 Oracle 12c 数据库中引入的多租用户环境（Multitenant Environment）中，允许一个数据库容器（CDB）承载多个可插拔数据库（PDB）。CDB 全称为 Container Database，即数据库容器，PDB 全称为 Pluggable Database，即可插拔数据库。在 Oracle 12c 之前，实例与数据库是一对一或多对一关系（RAC）：即一个实例只能与一个数据库相关联，而实例与数据库不可能是一对多的关系。当进入 Oracle 12c 后，可插拔数据库的引入，使实例与数据库可以是一对多的关系。

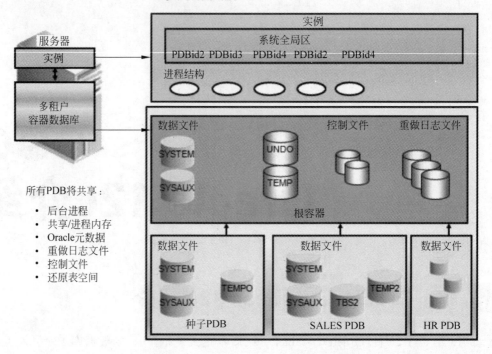

图 15-1

Pluggable Database 体系结构主要组成是由一个容器数据库(CDB)和多个可插拔数据库(PDB)组成,PDB 包含独立的 system 表空间和 sysaux 表空间等,所有 PDB 共享 CDB 的控制文件、日志文件和 undo 表空间。各个 PDB 之间互相是独立的,它们之间互相访问需要通过 DB Link。

15.3　Oracle 12c 新特性介绍

◆ Pluggable Databases 可插拔数据库

一个容器数据库(container database)中可以存放多个 Pluggable Databases。

(1) 对于外部应用程序和开发者来说,Pluggable Databases 看上去就是一个普通的版本 12.1 之前的数据库。

(2) 用户连接到 Pluggable Databases 时看到的是一个单一数据库。

◆ 新的管理模式

(1) 数据库管理员 DBA 可以连接到 Pluggable Database 并仅仅管理该数据库;

(2) 超级 DBA 可以连接到容器数据库并如同管理单系统镜像那样管理;

(3) RAC 中的每一个实例均打开容器数据库,并可以选择打开哪些 Pluggable Database。

◆ 内建的多分租(Multi-tenancy)

(1) 每个 Pluggable Database 均独立于其他 Pluggable Database;

(2) Resource Manager 特性被扩展到 Pluggable Database 中。

Pluggable Databases 特性可以带来的好处：

（1）加速重新部署现有的数据库到新的平台的速度；
（2）加速现有数据库打补丁和升级的速度；
（3）从原有的 DBA 的职责中分离部分责任到应用管理员；
（4）集中式管理多个数据库；
（5）提升 RAC 的扩展性和故障隔离；
（6）与 Oracle SQL Developer 和 Oracle Enterprise Manager 高度融合。

1. 创建一个新的 pdb

```
SQL> select * from v$version;

BANNER                                                              CON_ID
------------------------------------------------------------        ------
Oracle Database 12c Enterprise Edition Release 12.1.0.2.0 - 64bit Production   0
PL/SQL Release 12.1.0.2.0 - Production                                         0
CORE  12.1.0.2.0  Production                                                   0
TNS for Linux: Version 12.1.0.2.0 - Production                                 0
NLSRTL Version 12.1.0.2.0 - Production                                         0

SQL> select sys_context('userenv','con_name') from dual;
SYS_CONTEXT('USERENV','CON_NAME')
--------------------------------------------------------
CDB$ROOT

SQL> select con_id,dbid,con_uid,guid,name,open_mode,create_scn,total_size,block_size from v$pdbs;

CON_ID  DBID        CON_UID     GUID                              NAME      OPEN_MODE   CREATE_SCN  TOTAL_SIZE  BLOCK_SIZE
------  ----------  ----------  --------------------------------  --------  ----------  ----------  ----------  ----------
2       4264868981  4264868981  0951B9CFB39C2060E0530100007F6961  PDB$SEED  READ ONLY   1594434     870318080   8192
3       3977202490  3977202490  09521ABB982E7F0BE0530100007FD46A  XJNY      READ WRITE  1753354     917504000   8192

--创建一个新PDB
SQL> create pluggable database wapdb admin user wapadmin identified by wapadmin;

Pluggable database created.

SQL> select con_id,dbid,con_uid,guid,name,open_mode,create_scn,total_size,block_size from v$pdbs;

CON_ID  DBID        CON_UID     GUID                              NAME      OPEN_MODE   CREATE_SCN  TOTAL_SIZE  BLOCK_SIZE
------  ----------  ----------  --------------------------------  --------  ----------  ----------  ----------  ----------
2       4264868981  4264868981  0951B9CFB39C2060E0530100007F6961  PDB$SEED  READ ONLY   1594434     870318080   8192
3       3977202490  3977202490  09521ABB982E7F0BE0530100007FD46A  XJNY      READ WRITE  1753354     917504000   8192
4       4211777660  4211777660  0986C60CA76145FDE0530100007F304E  WAPDB     MOUNTED     1896905     0           8192
```

2. 克隆一个新的 pdb

```
SQL> create pluggable database mbs from wapdb;
create pluggable database mbs from wapdb
*
ERROR at line 1:
ORA-65139: Mismatch between XML metadata file and data file
/home/db/oracle/oradata/SMS/0986C60CA76145FDE0530100007F304E/datafile/o1_mf_system_b856b39h_.dbf for value of fcpsb (1897165 in the
plug XML file, 1904147 in the data file)


尝试将 wapdb 以 read only 模式打开
SQL> alter session set container = wapdb;

Session altered.

SQL> alter database open read only;

Database altered.

SQL> alter session set container = cdb$root;
```

```
Session altered.
SQL> select sys_context('userenv','con_name') from dual;

SYS_CONTEXT('USERENV','CON_NAME')
--------------------------------------------------------
CDB$ROOT

SQL> create pluggable database mbs from wapdb;

Pluggable database created.

SQL> alter pluggable database mbs open;

Warning: PDB altered with errors.

SQL> select con_id,dbid,con_uid,guid,name,open_mode,create_scn,total_size,block_size from v$pdbs;
```

CON_ID	DBID	CON_UID	GUID	NAME	OPEN_MODE	CREATE_SCN	TOTAL_SIZE	BLOCK_SIZE
2	4264868981	4264868981	0951B9CFB39C2060E0530100007F6961	PDB$SEED	READ ONLY	1594434	870318080	8192
3	3977202490	3977202490	09521ABB982E7F0BE0530100007FD46A	XJNY	READ WRITE	1753354	917504000	8192
4	4211777660	4211777660	0986C60CA76145FDE0530100007F304E	WAPDB	READ ONLY	1896905	870318080	8192
5	1281880106	1281880106	0987D2D2C15B4758E0530100007F4AA4	MBS	READ WRITE	1905674	891289600	8192

可见,在克隆 pdb 时,需要将源 pdb 以 read only 模式打开。

3. 将 pdb plug 到新的 cdb 中

1. 在源数据库创建 pdb 的 xml 文件
```
begin
    dbms_pdb.describe(
        pdb_descr_file => '/home/oracle/xjny.xml',
        pdb_name       => 'XJNY');
end;
/
```

2. 在目标数据库检查 pdb 的 xml 是否与目标库兼容
```
set serveroutput on
declare
    compatibleconstant varchar2(3) :=
        case dbms_pdb.check_plug_compatibility(
              pdb_descr_file => '/home/oracle/xjny.xml',
              pdb_name       => 'SALESPDB')
        when true then 'YES'
        else 'NO'
end;
begin
    dbms_output.put_line(compatible);
end;
/
```

3. 将 pdb 数据库 plug 到新的 cdb 中,在新 cdb 中执行
```
create pluggable database ebs using '/home/oracle/xjny.xml'
    nocopy
    tempfile reuse;
```

4. 打开新的 pdb
```
alter pluggable database ebs open;
```

4. 将 pdb 从 cdb 中 unplug

```
SQL> alter pluggable database ebs close;

Pluggable database altered.

SQL> alter pluggable database ebs unplug into '/home/oracle/ebs.xml';

Pluggable database altered.

SQL> select pdb_id,pdb_name,dbid,con_uid,status,con_id from dba_pdbs;

    PDB_ID PDB_NAME          DBID    CON_UID STATUS      CON_ID
---------- ---------- ---------- ---------- ---------- ----------
         3 XJNY       3977202490 3977202490 NORMAL              3
         2 PDB$SEED   4264868981 4264868981 NORMAL              2
         4 WAPDB      4211777660 4211777660 NORMAL              4
         5 MBS        1281880106 1281880106 NORMAL              5
         6 EBS        1440704945 1440704945 UNPLUGGED           6
```

如果不再需要 pdb，或者要将 pdb plug 到新的 cdb 中，需要 drop 掉 cdb

```
SQL> drop pluggable database ebs including datafiles;

Pluggable database dropped.

SQL> select pdb_id,pdb_name,dbid,con_uid,status,con_id from dba_pdbs;

    PDB_ID PDB_NAME          DBID    CON_UID STATUS      CON_ID
---------- ---------- ---------- ---------- ---------- ----------
         3 XJNY       3977202490 3977202490 NORMAL              3
         2 PDB$SEED   4264868981 4264868981 NORMAL              2
         4 WAPDB      4211777660 4211777660 NORMAL              4
         5 MBS        1281880106 1281880106 NORMAL              5
```

5. 在线重命名和重新定位活跃数据文件

不同于以往的版本，在 Oracle 数据库 12c R1 版本中对数据文件的迁移或重命名不再需要太多烦琐的步骤，即把表空间置为只读模式，接下来是对数据文件进行离线操作。在 12c R1 中，可以使用 alter database move datafile 这样的 SQL 语句对数据文件进行在线重命名和移动。而当此数据文件正在传输时，终端用户可以执行查询、DML 以及 DDL 方面的任务。另外，数据文件可以在存储设备间迁移，如从非 ASM 迁移至 ASM，反之亦然。

重命名数据文件：

```
SQL> alter database move datafile '/home/db/oracle/oradata/yingsu1/users01.dbf' to '/home/db/oracle/oradata/yingsu2/users_01.dbf';
```

从非 ASM 迁移数据文件至 ASM：

```
SQL> alter database move datafile '/home/db/oracle/oradata/yingsu/users_01.dbf' to '+DG_DATA';
```

将数据文件从一个 ASM 磁盘群组迁移至另一个 ASM 磁盘群组：

```
SQL> alter database move datafile '+data/yingsu/datafile/users_01.dbf' to '+DATA02';
```

在数据文件已存在于新路径的情况下，以相同的命名将其覆盖：

```
SQL> alter database move datafile '/home/db/oracle/oradata/yingsu/users_01.dbf' to '/home/db/oracle/oradata/yingsu2/users_01.dbf' reuse;
```

复制文件到一个新路径，同时在原路径下保留其拷贝：

```
SQL> alter database move datafile '/home/db/oracle/oradata/yingsu/users_01.dbf' to '/home/db/oracle/oradata/yingsu2/users_01.dbf' keep;
```

可以通过查询 v$session_longops 来监控这一过程。另外，也可以通过查看 alert 日志来查看。

6. 表分区或子分区的在线迁移

在 Oracle 12c 之前，如果需要在线移动分区表的分区到不同表空间，需要采用在线重定义的方式，操作步骤比较烦琐，在 Oracle 12c 中，可以采用在线 move 的方式来移动分区到新的分区表中。

```
SQL> alter table t_yingsu move partition|subpartition partition_name to tablespace tbs_yingsu;
SQL> alter table t_yingsu move partition|subpartition partition_name to tablespace tbs_yingsu update indexes online;
```

第一个命令是用来在 offline 状态下将一个表分区或子分区迁移至一个新的表空间。第二个命令通过加上 update indexes 和 online 命令，可以在线进行操作并维护表上任何本地或全局的索引，不阻塞其他 session 对该表的 DML 操作。

7. 不可见字段

在 Oracle 11g R1 中，Oracle 引入了不可见索引和虚拟字段，而在 Oracle 12c R1 中引入了不可见字段。在之前的版本中，为了隐藏重要的数据字段以避免在通用查询中显示，我们往往会创建一个视图来隐藏所需信息或应用某些安全条件。

在 12c 中，用户可以在表中创建不可见字段。当一个字段定义为不可见时，这一字段就不会出现在通用查询中，除非在 SQL 语句或条件中有显式地提及这一字段，或是在表定义中有 described。

```
SQL> create table t_yingsu(yno number(6), name name varchar2(40), sal number(9) invisible);
SQL> alter table t_yingsu modify (sal visible);
```

用户必须在 insert 语句中显式提及不可见字段名以将不可见字段插入到数据库中。虚拟字段和分区字段同样也可以定义为不可见类型。但临时表、外部表和集群表并不支持不可见字段。

8. DDL 日志

在 Oracle 12c 之前的版本中没有方法对 DDL 操作进行日志记录。而在 12c 中,用户可以将 DDL 操作写入 xml 和日志文件中。这对于了解谁在什么时间执行了 create 或 drop 之类的命令是十分有用的。要开启这一功能必须对 enable_ddl_logging 初始参数加以配置。这一参数可以在数据库或会话级加以设置。当此参数为启用状态,所有的 DDL 命令会记录在 $ORACLE_BASE/diag/rdbms/DBNAME/log|ddl 路径下的 xml 和日志文件中。一个 xml 中包含 DDL 命令, IP 地址,时间戳等信息。这可以帮助确定在什么时候对用户或表进行了删除或者是一条 DDL 语句在何时触发。

开启 DDL 日志功能:

```
SQL> alter system|session set enable_ddl_logging = true;
```

以下的 DDL 语句会记录在 xml 或日志文件中:

```
create|alter|drop|truncate table
drop user
create|alter|drop package|function|view|synonym|sequenc
```

9. 如何在 RMAN 中执行 SQL 语句

在 12c 中,用户可以在不需要 SQL 前缀的情况下在 RMAN 中执行任何 SQL 和 PL/SQL 命令,即用户可以从 RMAN 直接执行任何 SQL 和 PL/SQL 命令。如下:

```
RMAN> select username, machine from v$session;
RMAN> alter tablespace users add datafile size 500m;
```

10. RMAN 中的表恢复和分区恢复

Oracle 数据库备份主要分为两类:逻辑和物理备份。每种备份类型都有其自身的优缺点。在之前的版本中,利用现有物理备份来恢复表或分区是不可行的。为了恢复特定对象,逻辑备份是必需的。对于 12c,用户可以在发生 drop 或 truncate 的情况下从 RMAN 备份将一个特定的表或分区恢复到某个时间点或 SCN。

当通过 RMAN 发起一个表或分区恢复时,大概流程是这样的:

(1) 确定要恢复表或分区所需的备份集;
(2) 在一个临时辅助数据库中设置恢复到某个时间点;
(3) 利用数据泵将所需表或分区导出到一个 dumpfile;
(4) 用户可以从源数据库导入表或分区(可选);
(5) 在恢复过程中进行重命名操作。

以下是一个通过 RMAN 对表进行时间点恢复的示例(确保已经对稍早的数据库进行了完整备份):

```
RMAN> connect target "username/password assysbackup";
RMAN> recover table username.tablename until time 'TIMESTAMP…'
auxiliary destination '/home/db/oracle/oradata/yingsu/tablerecovery'
datapump destination '/home/db/oracle/oradata/yingsu /dpump'
dump file 'tablename.dmp'
notableimport -- this option avoids importing the table automatically.(此选项避免自动导入表)
```

```
remap table 'username.tablename': 'username.new_table_name'; -- can rename table with this option.
（此选项可以对表重命名）
```

注意事项：
- 确保对于辅助数据库在/u01 文件系统下有足够的可用空间，同时对数据泵文件也有同样保证。
- 必须要存在一份完整的数据库备份，或者至少是要有 system 相关的表空间备份。

在 RMAN 中应用表或分区恢复的限制和约束：
- sys 用户表或分区无法恢复；
- 存储于 sysaux 和 system 表空间下的表和分区无法恢复；
- 当 remap 选项用来恢复的表包含 not null 约束时，恢复此表是不可行的。

11. 临时 undo

在 Oracle 12c R1 之前，undo 记录是存储在 undo 表空间中的，在 12c R1 中引入了临时 undo 功能，那些临时 undo 记录现在就可以存储在临时表中，而不是存储在 undo 表空间中。临时 undo 的主要好处在于：由于 undo 记录不会写入 undo 日志，undo 表空间的开销减少并且产生的 undo 数据会更少。可以选择在会话级或数据库级开启临时 undo 功能。

要使用这一新功能，需要做以下设置：

（1）兼容性参数必须设置为 12.0.0 或更高；

（2）启用 temp_undo_enabled 初始化参数；

（3）由于临时 undo 记录现在是存储在一个临时表空间中的，需要有足够的空间来创建这一临时表空间；

（4）对于会话级，可以使用 alter system set temp_undo_enable=true。

以下所列的字典视图是用来查看或查询临时 undo 数据相关统计信息的：

- v$tempundostat
- dba_hist_undostat
- v$undostat

要禁用此功能，只需做以下设置：

```
SQL> alter system|session set temp_undo_enabled = false;
```

12. 限制 PGA 的大小

在 Oracle 12c R1 之前，对单个 session 使用的 PGA 大小没有公开的参数来硬性限制。当设置 pga_aggregate_target 参数后，Oracle 会根据工作负载和需求来动态地增大或减小 PGA 的大小。而在 12c 中，可以通过开启自动 PGA 管理来对 PGA 设定硬性限制，这需要对 pga_aggregate_limit 参数进行设置。通过该参数来对 PGA 设定硬性限制以避免过度使用 PGA。

```
SQL> alter system set pga_aggregate_limit = 2g;
SQL> alter system set pga_aggregate_limit = 0; -- disables the hard limit
```

重要提示： 当超过了当前 PGA 的限制，Oracle 会自动终止/中止会话或进程以保持最合适的 PGA 内存。